COLLINS GEM
ANTIQUE MARKS

GEM

COLLINS GEM
CRICKET

COLLINS GEM
DIETING

COLLINS GEM
DOGS

COLLINS GEM
FIRST AID

COLLINS GEM
INTERNET

COLLINS GEM
PREDICTING

COLLINS GEM
Ready REFERENCE

COLLINS GEM
SHARKS

COLLINS GEM
WHALES & DOLPHINS

COLLINS GEM
WHISKY

COLLINS GEM
WORD PROCESSING

COLLINS GEM
Your PC

COLLINS GEM

TIME

An Exploration of our Discovery
and Experience of Time

N. M. Wells

Consultants:
Dr James Mackay
Dr R. N. F. Walker

HarperCollins*Publishers*

Author: N.M. Wells has created a great many works on historical, religious, philosophical and arts-oriented subjects. His most recent book was on William Morris.

Consultant: Dr R.N.F. Walker is a dedicated astrophysicist; now lecturing at Bristol University he has published and broadcast extensively in his specialist area.

Consultant: Dr James Mackay, a journalist, broadcaster, biographer and historian, has edited and contributed to a large number of general and specialist encyclopedias.

Grateful thanks also for the patience and assistance of Jonathon Betts, Curator of Horology at the Royal Greenwich Observatory.

HarperCollins Publishers
PO Box, Glasgow G4 0NB

First published 1999

Reprint 10 9 8 7 6 5 4 3 2 1 0

ISBN 0 00 472472-0

Created and produced by Flame Tree Publishing, part of The Foundry Creative Media Co. Ltd
Crabtree Hall, Crabtree Lane, London SW6 6TY

Printed in Italy by Amadeus S.p.A.

Contents

Seventeenth-century celestial map showing the constellations and the planets of our solar system.

Introduction

FOR MANY people time is a simple matter of days, hours and seconds. It determines the pattern of their lives and is taken as a matter of objective reality. For others, time is a magical notion, subtle and shifting, a fugitive spell that washes over everything and everybody. Cosmologists, poets, painters and geologists all appear in this latter category, making even the most fanciful explorations or dalliance with time worthy of respect.

The subject of time is clearly too vast to cover in detail in such a short reference volume as a Gem, but it is possible to convey the flavour and passion of its multi-faceted fascination so as to encourage further reading in any particular area of interest uncovered here.

There has been a number of earnest attempts at popularising aspects of the history of time, notably by mathematical physicist Stephen Hawking in *A Brief History of Time*. The exceptional success of Hawking's work suggests an inherent fascination with the subject, even if many people buy such books to dress their coffee tables. It can be argued that by engaging with the subject at all we challenge the passive dominion of time and empower our lives. Of course, business books are full of time-management schemes, designed to increase effectiveness at work by reducing the size of the in-tray,

Observing the heavens with astrolabes and other early instruments, thirteenth century.

delegating and squeezing an extra minute or so from each hour. This book should show that there is more to time than merely playing the game in which time makes us its slave because historically and collectively we agree that this is so.

Calendars, clocks, behaviour (both sociological and psychological) and religion are all touched on in the following chapters, but the book concludes by delving into space and time. You might regard the rarefied theories of, for instance, time as a fourth dimension or hyperspherical space-time with some

Celestial globe, 1625.

amusement, because they are so far from our everyday experiences, but you will also discover that the almost maddening tendency of scientists to try to outwit themselves, to confirm both their own and their colleagues' theories, speaks of a spirit and humanity not found in some of our best literature.

Finally, it is worth touching on a sensitive subject: is there a place for God in this exploration? Is He central, implicit or beyond our scope? Science is about the constant search for definition and truths, so that while some religions revel in the mystery of their God, science always – and simply – wants to find out more. This book does discuss some of the issues regarding the existence and nature of God, in or outside time, and points to some areas of further discovery.

Time Now

IN MODERN western cultures we experience time as a straightforward sequence of seconds through to years. Every year new calendars and diaries are published containing data supplied by government and religious bodies responsible for regulating holiday and festival dates. Newspapers are often used as confirmation of the day of the week, while we rely on our watches and the radio for more precise references of time.

Humankind has taken over seventeen thousand years to accumulate the knowledge to manage time in this way and the story of its progress and the exploration of its implications can be found in the following pages.

DAYS OF THE WEEK

AS WE WILL see later, we have based our concepts of the year, the month and the day on the observable motions of the stars, the sun and the moon and most recently on the oscillations of an atomic particle (see p. 157). However, the number of days in a week is entirely man-made and has its origins in Roman

---

A TIME TABLE

- Each year is 365 days long.
- Each year is divided into four seasons:
  Spring, Summer, Autumn, Winter.
- Each year has 52 weeks.
- Each week has seven days.
- Each week has 12 months.
- Eleven months have 30 or 31 days.
- February has 28 days, 29 days in leap years.

THE MILLENNIUM?

We live at the edge of the third millennium, 2000 years after a date nominated as the birth of Christ (see p. 32). There are, however, a number of disputes about the actual date of the millennium and the actual birthdate of Christ; in addition, many other cultures have a different number for the year which Europe and the West calls 2000:

- Muslim calendar: 1420
- Buddhist calendar: 2544
- Mayan Great Cycle: 5119
- Jewish calendar: 5760
- Ancient Egyptian calendar: 6236
- Chinese calendar: Year of the Dragon

and Christian attempts at maintaining order throughout their fields of influence. A week of seven days is, however, universally applied, used by almost all of the world's cultures including Chinese, Japanese and African societies.

It originated during the pre-Christian Roman period, when Julius Caesar controlled or had conquered the whole of Europe, the Middle East, parts of Asia and North Africa. The

- Each day runs 24 hours from midnight to the following midnight.
- Different parts of the world have different daylight hours, coordinated by international agreement between governments.
- Each hour contains 60 minutes.
- Each minute contains 60 seconds.
- Each second can be broken down further into tenths, hundredths, thousands and millionths of a second.

Romans inherited the variety and richness of knowledge
accumulated by the Egyptians, Persians, Greeks, Jews and,
through extensive trade, Chinese. Babylonian and Sumerian
number and hieroglyphic systems also influenced these
cultures, and through them we have retained the unit of 60
for our minutes to hours and seconds to minutes.

Astronomy and mathematics were particular pre-
occupations of each of these ancient civilizations and it was a
natural development in the centuries either side of Christ's
birth to think of defining the week by using the names of
planets and the sun and moon, all long observed and used as
the basis of calculations of other elements of time, such as the

Planets	Saturn	Sun	Moon
Languages			
Czech	sobota	neděle	pondělí
Danish	Lørdag	Søndag	Mandag
Dutch	Zaterdag	Zondag	Maandag
English	Saturday	Sunday	Monday
French	samedi	dimanche	lundi
German	Samstag	Sonntag	Montag
Hebrew	shabat	yom rishon	yom sheni
Hungarian	szombat	vasárnap	hétfö
Italian	sabato	domenica	lunedí
Norwegian	lørdag	søndag	mandag
Old Latin	Saturni	Solis	Lunae
Old Saxon	Saterne's	Sun's	Moon's
Polish	sobota	niedziela	poniedzłarek
Romanian	simbata	duminica	luni
Spanish	sabado	domingo	lunes
Russian	subbota	voskresenie	ponedelnik
Serbo-Croat	Subota	Nedjelja	Ponedjeljak
Swedish	lördag	söndag	måndag

extent of the year. In AD 325 Emperor Constantine made Christianity the religion of the Roman Empire, so incorporating the pagan, planet descriptions of the week into the newly organised religion. The names survive today in most languages, as can be seen below, but the exceptions, fully endorsed by the Catholic Church of the time, include other pagan descriptions such as the Nordic influence on the English names of the week (see p. 34).

Before medieval times, the planets and the sun, were thought to revolve around the earth.

Mars	Mercury	Jupiter	Venus
úterý	středa	čtvrtek	pátek
Tirsdag	Onsdag	Torsdag	Fredag
Dinsdag	Woensdag	Donderdag	Vrijdag
Tuesday	Wednesday	Thursday	Friday
mardi	mercredi	jeudi	vendredi
Dienstag	Mittwoch	Donnerstag	Freitag
yom shlishi	yom Revi'i	yom chamishi	yom shishi
kedd	szerda	csütörtök	péntek
martedí	mercoledí	giovedí	venerdí
tirsdag	onsdag	torsdag	fredag
Martis	Mercurii	Jovis	Veneris
Tiw's	Woden's	Thor's	Frigg's
wtorek	środa	czwartek	piątek
marti	miercuri	joi	vineri
martes	miércoles	jueves	viernes
vtornik	sreda	chetverg	pyatnitsa
Utorak	Sreda	Četvrtak	Petak
tisdag	onsdag	torsdag	fredag

TIME AND THE CALENDAR

Introduction: Beginnings

*In the time of dreams, two brothers, the Bagadjimbiri,
come out of the earth as dingoes. They turn into two
giants in human form and grow so tall that their heads
touch the sky. Nothing exists before this: no trees, no
animals, no people. The brothers come out of the earth
just before the dawn of the first day. A few moments
later they hear the call of a little bird, the duru, which
always sings at the break of day, so they know it must
be dawn. They see plants and animals and give them
names. Once named, they come to life.*

THIS DREAMTIME creation myth from the Karadjeri
Aboriginals of Australia is retold, orally, always in the same
form, with the storyteller connecting to and being part of this
beginning of time, at once inside the past and a part of the
present. The Karadjeri believe that at the point of creation a
fundamental rhythm is established, the call of the little bird at
daybreak, so that a definition of time is imposed on humankind by the natural order of things.

ORIGINS OF THE MODERN CALENDAR

THE HISTORY of humankind's measurements of days,
months and years is the history of civilisation and, it has been
argued, its titanic conflict between sacred and secular political
forces. It has propelled us through a history of ideas and

understanding to a point at the start of the third millennium where we can measure the beginnings of the universe itself, rather than just rely on articles of faith, myth or simple conjecture.

The Ancient World, with its immense reserves of knowledge stretching from the ancient Chinese to the Greeks, Babylonians and Vedic Indians, struggled to explain the rhythm of the year using mathematical and astronomical observations that were, by the end of the first millennium, the envy of an embarrassed Europe cloaked in the Dark Ages. The balance of knowledge then shifted, with ideas and learning flooding through Europe, resulting in the powerful cultural synthesis of the Renaissance and ultimately causing the separation of sacred and secular authority over the instruments of time. This is significant because religion and what might be termed organised superstition had played the primary role in controlling the lives and minds of humankind for as long as the skies and the stars had been relied on for succour and inspiration.

Our view of the universe has moved from earth-centred (geocentric) systems to a more humble appreciation of our place beside the millions of stars and planets, and events like exploding black holes.

Definitions of the Astronomical Year

There are many different types of year, and the differences fuel the twists and turns in the story of time and the calendar.

THE LENGTH of the year depends on where you measure it from. Our perceptions of time have changed and developed as science and technology have improved the accuracy of its measurement over the past 4000 years.

LUNAR YEAR

THIS IS the most easily observable type of year and it formed the basis of many cultures' early government of time. As each month starts at the new moon and lasts for 29.5306 days this was generally translated into alternating months of 29 days (hollow months) and 30 days (full months). This gives a year of 354 days: too short by 0.3672 of a day of the true lunar year, so in order to redress the balance, a leap day every third year had to be inserted. (The addition of days or other units, like seconds, to balance the calendar, is known as an intercalation.) To make this calculation more accurate, a further leap day would have to be added every 10 years.

Over time, lunar years become out of sync with the seasons which are determined by the solar calendar. Several cultures, including the ancient Chinese and Greeks, used a combination of lunar and solar years by adding leap months and combining the seasonal cycles with the moon's phases (the time from one

new moon to the next), creating in one common solution a
19-year solar cycle which coincided with 235 lunar months.

SOLAR YEAR

THERE ARE a number of recognised forms of the solar year,
all necessary because of the slightly elliptical rotation of the
earth around the sun:

- The tropical year – marks the year of the sun's passage
  between two vernal (spring) equinoxes. The seasons are
  fixed and the length of the year is 365.242199 days.
- The sidereal year – marks the passage of the sun relative
  to a fixed star. The length of the year is 365.256366 days.
- The anomalistic year – marks the passage of the sun at
  the point of the perihelion, when the earth is nearest to
  the sun. The length of the year is 365.259636 days.

The tropical year is used where the solar year is mentioned.

**The earth's elliptical orbit around the sun, showing the Perihelion
for AD 2000, with dates for the northern hemisphere.**

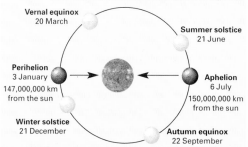

Vernal equinox
20 March

Summer solstice
21 June

Perihelion
3 January
147,000,000 km
from the sun

Aphelion
6 July
150,000,000 km
from the sun

Winter solstice
21 December

Autumn equinox
22 September

The Use of Different Calendars

The significant differences in the length of the solar and lunar years and the enduring reliance on both the sun and the moon to determine the daily routines of life reveals some fundamental paradoxes which needed to be controlled by a central calendar-making authority.

T HE DEVELOPMENT of our record of everyday time, as expressed through the use of calendars, is a vivid story of tension and conflict: between religious institutions and political authorities, between factions within faiths and, fundamentally, between the different lengths of the solar year and the lunar year.

Western Christian faiths valued the date of Easter above all others. As most cultures were strictly ruled either by or alongside a religious authority, the dates of worship were more than a simple matter of noting time. Like the sculpting of the hidden gargoyles of Canterbury Cathedral (the actual making itself was the worship), or the rituals of Shabat to Jews, certain actions are an article of faith in their own right, whether observed publicly or in private: a fundamental expression of a whole way of life.

THE EFFECT ON DAILY LIFE

FOR MOST normal people, up to the end of the Middle Ages, life was lonely, dirty and extremely hard work. The rhythm of the seasons and the ending and arrival of daylight were the pillars of their short lives (people lived on average no longer than 35 years).

Unfortunately, these two rhythms – the season and the day – work against each other, because the 12 lunar months of 29½ days produce a shorter year than the measurement from one year to the next of the date of the spring and autumn equinoxes

Medieval shepherd (1504) orients himself by the simplest of methods: using his eyes to observe the stars.

MEAN STARTING DATE FOR EACH SEASON		
Vernal equinox	**equal hours, day and night**	21 March
Summer solstice	**longest day of the year**	22 June
Autumn equinox	**equal hours, day and night**	23 September
Winter solstice	**shortest day of the year**	22 December

(when the sun crosses the equator and day and night are of equal length everywhere on earth, generally around 21 March and 23 September) and the summer and winter solstices (22 June and 22 December). The lunar year, at 354 days, is just over 11 days short of the solar year. A further complication is that, because the earth does not rotate around the sun in a perfect circle, there is some shifting between one year and the next for these equinoxes and solstices.

GLOBAL CONSIDERATIONS

PRACTICAL PROBLEMS only manifest themselves over many years – for example, the Julian calendar became a concern for the Christian faith, which placed high authority on the date of the risen Christ at Easter, only to realise that the date had moved backwards. By the time of the Gregorian

Christ's return from the grave is the reason for the importance of Easter to most Christians.

calendar reform in the sixteenth century, Easter was 11 days nearer summer, a long way from the spring equinox. In AD 325 the Council of Nicea (see p. 29) had agreed that Easter should be on the first Sunday after the first full moon after the vernal equinox, making it between 23 March and 25 April. Over the centuries there have been many attempts to fix Easter, like Christmas, to a particular date, one of the most recent of which was a move in 1923 by the then League of Nations.

A further practical issue is the need to establish a reliable system between cultures which trade or war with each other. A recent, unfortunate example was the NATO bombing of Iraq in late 1998, a campaign that intruded into Ramadan against the explicit intentions of the Alliance, who were attempting to retain some

In Pakistan, muezzin calls the faithful to prayer during the day.

moral position by ceasing the attacks during this holy Muslim time. The calendar used by the Iraqis is slightly different to the one generally accepted in Europe and the USA, a fact not understood by the Alliance Intelligence services, with disastrous human and political results.

Ancient Calendars

Every significant ancient culture made astonishingly
accurate measurements of celestial motion, but they
all had a range of solutions to the problem of the
variance in the lunar and solar year.

LUNAR CALENDARS

TWENTY THOUSAND years ago, Ice Age hunters scratched
lines and holes in bones to show the days before the next new
moon. The discovery of an eagle bone at le Placard, dated to
13,000 BC, is one of the earliest examples of man recording
himself against a natural phenomenon. Similar observations
were made by the descendants of all the major ancient
civilisations, but as they approached what is now known as
6000 BC and 5000 BC, their increasing knowledge of astronomy led them to note the differences between lunar and
solar time and to find ways of periodically catching up.

CULTURES OF THE TIGRIS AND EUPHRATES RIVERS

SOME WRITTEN records exist from these ancient cultures,
although much has been lost, particularly about the
Babylonians (covering part of what is now Iraq) – whose
learning appears to have been particularly superior – due to the
great fires at the libraries in Alexandria of 97 BC and AD 696.
We know that their lunar year, before 2000 BC, consisted of 12
months of alternating 29 and 30 days. The Sumerians, their
successors in many ways, rounded the lunar month up to 30

*The Tower of Babel, although unfinished, was a celebrated and
breathtaking product of the learning and skill of the Babylonians.*

days, giving them 360 days every year, which neatly dovetailed with their base 60 numbering system, forming the basis of the imperial numbering system which gave us 12 inches in a foot and 24 hours in a day.

SOLAR YEARS

SOME EVIDENCE of time notation remains in the form of artifacts. Stonehenge, for instance, is an incredible formation of standing stones which, whatever its other purposes, perfectly records the summer solstice, designed so the sun shines down the central avenue, illuminating the centre stone.

Archeological finds and early documents have revealed that the ancient Greeks added 90 days every eight years to make up the difference between the lunar and solar years. The Chinese, by *c.* 2350 BC, had added seven months every 19 years. Jewish astronomers added one month every three years with a further month added by decree when necessary.

Unusually, the Mayans of Central America used a solar year of 365 days, with 18 months of 20 days each and a further five special days. In fact, they also had another calendar of 260 days, which was the cycle reserved for days of worship. Using this – combined with the 365-day calendar – they calculated what they termed Calendar Rounds. Separately they measured Long Cycles of 360-day units called *tuns*, which they multiplied by 400 then 13 to create their Great Cycles of 5130 years which marked the end of one life and the start of another. The end of the current cycle, the end of the world according to Mayan predictions, is AD 2012.

THE ANCIENT EGYPTIANS

THE ANCIENT Egyptians are especially important to our concept of time because their intellectual and cultural influence had a major impact on the development of the Julian calendar, leading to the Gregorian form we use today.

The Nile, as the source of life, was also the key to their understanding and observations of time. Using a nilometer, possibly as early as 5000 BC, they determined that the year was 365 days long. The nilometer had notches or steps in the banks of the Nile marking the different levels of water height from the low point in May to its highest in September, so that they could plan for the floods between June and October, the planting and growth of crops from October to February and the harvest from February to June. The shadows cast by the pyramids were also used to measure equinoxes, in a similar way to the menhirs of Stonehenge.

The ancient Egyptians' advanced pursuit of knowledge led them to observe that the star we know as Sirius rose in line with

Great Cycle (Baktun x 13)
1,872,000 days. 5130 years.
Baktun (Katun x 20)
144,000 days.
Katun (Tun x 20)
7200 days.
Tun
360 days.

The Mayan Great Cycle was their longest calendar period. Parts of the Mayan calendar were included in Aztec calendar stones (right).

the sun at the point every year when the annual inundation of the Nile occurred. This enabled them to conclude that the year was in fact 365¼ days long.

In a move which has echoes throughout the history of time, however, local priests resisted this slight change, having adopted the 365-day year as sacred, even though scientific observation was proving them to be incorrect. The priests saw science as undermining their authority. We will repeatedly return to this theme throughout Europe's Dark Ages.

ALEXANDRIA

BY 334 BC, Alexander the Great had conquered Greece, Egypt and Persia – three of the most literate, philosophically and scientifically inclined nations in the world. The knowledge from these three civilisations was brought together in the cosmopolitan Egyptian city of Alexandria, built by Alexander in 332 BC. This city, with its population of 300,000 (not counting slaves) by the first century AD, also housed the most extensive library of the world's great literature and learning, including Aristotle's works. Waterclocks were created here, an encyclopedia of astronomy was drawn up, mathematical treatises and musings on the nature of time and man were distributed.

It is possible to trace the origins of all core principles behind the

Alexander the Great.

In the second century AD, Ptolemy lived and studied in Alexandria. His famous summaries of Greek astronomical ideas and his theories of the universe dominated western ideas on astronomy until the time of Copernicus. He placed earth at the centre of the universe, with the planets revolving around it. Behind him is the goddess Astronomia.

measurement of time and the calendar to the astronomical and mathematical scholarship of Alexandria in the two centuries before the birth of Christ and beyond.

CAESAR AND CLEOPATRA

SHAKESPEARE'S DRAMATIC Cleopatra seems to have been an accurate portrayal of the intelligent, passionate, politically sagacious historic figure who seduced Julius Caesar in order to remove her brother Ptolemy from the stewardship of an Egypt which had by then been conquered by the Roman Empire.

Caesar was an immensely powerful military and political leader, whose success had given him power over the whole of Europe, North Africa and the Middle East: more than half the known world of the time. The chaos and dispute of dates throughout the Roman Empire made Caesar determined to bring the calendar under his authority: it is said that in 47 BC, at a feast arranged by Cleopatra to honour Caesar, he discussed the Egyptian method of measuring the year with the celebrated astronomer Sosigenes and decided how to effect a far-reaching reform of time.

The Julian Calendar

The Julian calendar was altered and misunderstood, but ultimately became the basis for all subsequent records of time until 1582.

UP TO THE early years of Caesar, the Romans calculated the years against the date of the founding of Rome in 753 BC, or the reign of the succession of emperors. *Calends* (from which the modern word 'calendar' is derived), meaning the coming together of people, was the first day of the month, a time when the priests would announce the sacred events and festivals of the coming month.

The Julian calendar removed the 354-day lunar year calendar, introducing a 365-day solar version, with leap years every four and alternated 30- and 31-day months, with 29 days in February. To bring it into line with the observable motion of the stars, 69 days had to be added to the first year; called the Year of Confusion, this lasted 445 days, with the first day of January starting in the old month of March.

Julius Caesar, 100–44 BC.

Caesar's reforms were an important part of the struggle between the sacred and the secular, relieving the priests of their central role in defining the calendar. The objective fact of the day was, for the first time, not subject to the machinations of the priests and their sponsored politicians.

AUGUSTUS CAESAR

THE PROMISE of stability was undermined initially when the leap years were applied every three years instead of every four. This was eventually spotted by Emperor Augustus, so that

from AD 4 the dates were correct. Augustus indirectly caused the other main problem in the early years: in recognition of Caesar's work on the calendar, the senate had changed the name of the month Quintillius to Julius, which we know as July. Similarly, because Augustus had completed a stunning series of military victories in AD 8, so the month of Sextilius was changed to Augustus. Unfortunately this month had only 30 days to Julius's 31, so they took one day from February and reorganised the 30- and 31-day months from September, producing the disorganised second half of the year which persists today.

MONTHS

The English names for the months have their origins in the Latin language and come to us from the Roman Empire.

JANUARY: *Januarius.* After the god Janus.

FEBRUARY: *Februarius. Februa* was the Roman festival of purification.

MARCH: *Martius.* After the God Mars.

APRIL: *Aprilis.* Either from the Greek god Aphrodite, or the Latin *aperire*, meaning to open, April being the opening month of spring.

MAY: *Maius.* After the goddess Maia.

JUNE: *Junius.* After the goddess Juno.

JULY: *Julius.* After Julius Caesar, in 44 BC, formerly *Quintillius*, from the Latin for fifth, *quintus*, being the fifth month of the old Roman calendar.

AUGUST: *Augustus.* After Emperor Augustus, in 8 BC, formerly *Sextilius*, from the Latin for sixth, *sextus*, being the sixth month of the old Roman calendar.

SEPTEMBER: *September.* From the Latin for seven, *septem*, being the seventh month of the old Roman calendar.

OCTOBER: *October.* From the Latin for eight, *octo*, being the eighth month of the old Roman calendar.

NOVEMBER: *November.* From the Latin *novem*, nine, being the ninth month of the old Roman calendar.

DECEMBER: *December.* From the Latin for ten, *decem*, being the tenth month of the old Roman calendar.

CONSTANTINE THE GREAT

IN AD 312 Constantine moved the centre of the Roman Empire from Rome to Byzantium, renaming it Constantinople. With its closer proximity to Alexandria, Byzantium was a thriving cultural centre and afforded greater prestige than Rome, which at that time was suffering the fatigue of empire and leadership. The origins of the Eastern Orthodox Church and the great schism with Rome can be found in this move.

Constantine declared himself a Christian and, as a Christian ruler, proceeded to impose his will throughout the Roman Empire, reversing the secular authority established by Caesar over 300 years earlier. He called a gathering of all bishops to Nicea in Turkey to resolve the many differences of faith and modes of worship between the sects that had developed throughout the years since the death of Christ. One main issue was the date of Easter.

After much painful and difficult debate, Constantine was able to make three significant alterations to the Julian calendar. Sunday, at the end of the seven-day weekly cycle, was prescribed as a holy day, deliberately not the Saturday Sabbath of Judaism. Christmas was a fixed holiday, while Easter was the first Sunday after the first full moon after the measurably moveable spring equinox. It is said that the phrase 'moveable feast' originates from this feast of Easter. The first council of Christian bishops issued what became known as the Nicene Creed, laying the foundation stones for a single, unified Christian faith, the Catholic (from the Greek *katholikos*, 'universal') Church.

The Dark Ages of the West

The end of the Roman Empire plunged its territories into a miserable period of ignorance and chaos, which left the accumulation of knowledge to the East and Far Eastern countries. Time was left to its own devices as the people struggled to survive; natural philosophy (the discipline that would later become known as 'science') was regarded as heretical.

THE END OF THE EMPIRE

FROM *c.* AD 350, Rome was cursed by internal revolts and threats from beyond its territories. In AD 410 Visigoths smashed through the heart of the Empire and terminated one of the most effective, productive and ordered civilisations by the sacking of Rome. By *c.* AD 450 Europe was a bloody pulp of war and invasion, with Aryan barbarian hordes, the Franks, Berbers, Lombards and Ostragoths, breaking into old Roman territories, looting everything they could find and then fighting one other.

Inevitably the appetite for intellectual discovery, including the development of astronomy and its use for the calendar, was much reduced, so the peasantry, and most of the monasteries, sank back to their old reliance on the

Saint Augustine in his cell.

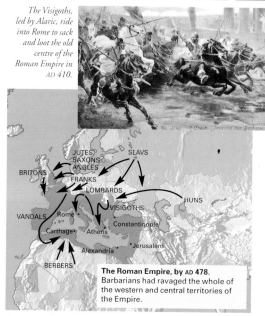

The Visigoths, led by Alaric, ride into Rome to sack and loot the old centre of the Roman Empire in AD 410.

JUTES
SAXONS
ANGLES
SLAVS
BRITONS
FRANKS
LOMBARDS
VISIGOTHS
HUNS
VANDALS
Rome
Carthage
Constantinople
Athens
Alexandria
Jerusalem
BERBERS

The Roman Empire, by AD 478.
Barbarians had ravaged the whole of the western and central territories of the Empire.

lunar months. Scientific matters, due to a misunderstanding about the teachings of the influential Christian philosopher, Augustine of Hippo (AD 354–430), were regarded as potentially heretical because such matters as time were felt to be in the hands of God.

THE CATHOLIC CHURCH INHERITS
THE ROMAN EMPIRE

CONSTANTINE'S NICEAN Council, in cementing the authority of the Christian Church, enabled the continuation of the structures of the Roman Empire through the Catholics. The Barbarians wanted land and money, but they had respect for the spiritual authority and dignity of the Roman Bishopric to the extent that they were converted over the following two centuries to the Catholic faith.

In AD 525, Pope Hilary asked mathematician and astronomer Dionysius Exiguus (AD 500–560) to calculate the next 95 years of Easter dates. This was part of the gathering conflict with the Alexandrian-backed eastern Christian Church, which had always been able to provide the astronomical knowledge for such tasks, unimpeded by the chaos of the European Dark Ages.

Dionysius's work was the first to use Anno Domini dates based on the birth of Christ. Unfortunately, we now know

---

COUNTING THE YEARS

All dates are accompanied by abbreviations, the meanings of which are sometimes obscure.

Abbr.	Meaning	Origin
AUC	*ab urbe condita*	Latin for 'from the founding of the city' (Rome, 753 BC), used before the adoption of AD.
AD	*Anno domini*	Latin for 'in the year of our Lord', i.e. the birth of Christ.
BC	Before Christ	Used since the eighteenth century.
CE	Common Era }	Both used to avoid reference to the
BCE	Before CE }	overtly Christian dating system of AD and BC.
AH	After *Hidjra*	Used by Muslim writers.

The Venerable Bede.

that, as Jesus must have been born while King Herod was still alive, the year of his birth is more likely to be between 6 and 4 BC (the year of Herod's death), but Dionysius's work was, nevertheless, a sign of progress during the depths of the Dark Ages. He had made his changes using the best-available knowledge, being careful to couch the terms of his research as being in the spirit of God, as an article of faith, rather than emphasising empirical evidence. An English monk, the Venerable Bede (AD 673–735), wrote the only surviving history of the Dark Ages, and used Dionysius's Easter dates and conventions.

By the early 700s, events in the Middle East further strengthened the Catholic Church as Muslim armies conquered much of the Byzantine lands, allowing Rome to exert its independence from the cultural powerhouse of Constantinople. It should be remembered that the Roman Church could assert superiority because of the strong tradition that Christ's disciple Peter founded the Church there and became the first Bishop of Rome.

The barbarians, having destroyed the Roman Empire, eventually settled and converted to Christianity.

THE JULIAN CALENDAR AND BRITAIN

IN AD 664, at the Synod of Whitby, the two main Christian traditions in the British Isles – the Celtic and Roman

The Norse god Thor.

Christians – agreed to adopt the Julian calendar and its dates for Easter.

In order to placate the dominant Saxons, who had invaded then settled in Britain, the Christian Church allowed the Saxon's gods to be used for the weekdays, in a way which did not interfere with the integrity of the calendar or the Easter dates. To this day the English-speaking world retains Tiw's day, Woden's day, Thor's day and Freya's day (even the goddess

Eostre's name was adopted for Easter), while much of the rest of Europe kept the Roman forms (see p. 10).

CHARLEMAGNE

AS THE influence of the Catholic Church spread, so did the endless bickering and factional feuding within Rome and between its senior bishops. Pope Leo III was blinded and his tongue torn out by his enemies. Charlemagne (AD 742–814), the King of the Franks, rushed to his aid and for his services he was made the first Holy Roman Emperor on Christmas Day AD 800.

This significantly strengthened the grip of the Roman Church because it now had temporal protection in the form of the armies of Charlemagne. This balance introduced a stability that lasted for centuries and ushered in a renewal which led Europe out of the Dark Ages. Indeed, Charlemagne was a barbarian king who had the zeal of a convert, and he became a champion of knowledge, the arts and sciences. The Caliph of Baghdad, Haroun al-Rashid ('Aaron the Wise'), the master of the Islamic world, sent a remarkable gift of a brilliantly intricate, decorated and crafted clock which would strike on the hour. Charlemagne is said to have been ashamed that he and his people could not match the splendour and learning this object represented and he made great efforts to encourage scholarship throughout his Catholic, European world. This was the time of the Carolingian Renaissance, characterised by the exchange of ideas with the inheritance of the Alexandrian treasure houses of knowledge.

Charlemagne (right), ruler of France and Holy Roman Emperor, with Pope Leo III (left) and St Peter (centre).

NUMBERS AND FRACTIONS

TWO OTHER areas which highlighted the extent of Western ignorance were a lack of fractions smaller than a quarter and the Hindu numbering system of one sequentially through to nine. The former was critical in solving the timing problem inherent in the Julian calendar because the actual solar rotation is slightly less than the quarter day over the 365. Because the West did not have the mathematics (or indeed the religious flexibility) to quantify this, the calendar could not be corrected with any accuracy. The numbering system that we use today was developed in the Vedic period in India, 2000 BC, as a means of expressing large numbers concisely. Zero was not recognised as a number until the end of the first millennium.

TRADE, EMBARRASSMENT AND KNOWLEDGE

A NUMBER of events occurred around the time of the first millennium which had some effect on the development of the calendar.

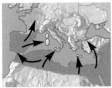

By AD 763 the whole of North Africa, Egypt, Syria, Persia, Spain and Portugal was held under Arabic rule, with constant attacks on the heart of Europe.

Since the creation of the Islamic religion in AD 622 (the first year of the Muslim calendar), Muslim armies had conquered much of North Africa, the Middle East and parts of Europe such as Spain and Portugal. Unlike the Aryan barbarian hordes which had smashed their way through the old Roman Europe, the Islamic invaders brought the ideas and learning of their own more sophisticated culture and integrated themselves with the conquered societies. From AD 900 this, and the influx of trade from the East, now that the European countries had stopped fighting amongst themselves, led to a rekindling of the explorations which had been halted by the decline of the Roman Empire.

Medieval Arabic painting of Muhammad's vision which led him to establish the new faith of Islam.

The power of the pope – the Holy Roman Emperor Heinrich IV does penance before being forgiven by Pope Gregorius VII.

By 1100, Catholicism was at the height of its power, having become the dominant Christian belief system throughout the West. It also resolved its long-running disputes with the

Eastern orthodox churches by excommunicating its primary bishop, the Patriarch of Constantinople, in 1054, for not recognising the supreme authority of the Pope. The Catholic and Eastern orthodox churches split irrevocably.

FAILED ATTEMPTS

CONTACT WITH other, more sophisticated cultures revealed the inadequacies of the Julian calendar, but the culture of the West was still held back by the principle that scientific matters, such as the motion of the stars and the accurate measurement of time, were the gift of God and it was almost heretical to try to be God-like by measuring and predicting motion.

In 1277, Roger Bacon (*c.* 1214–92), an English Franciscan monk, pointed out to Pope Clement IV that the calendar was disastrously wrong, so much so that the Easter and Lent dates were fundamentally and obviously inaccurate. He calculated that an extra 11 minutes in the Julian compared to the true solar calendar had resulted in a shift of one day every 125 years. An idiosyncratic but fiercely scholarly man, his incredible scientific labours went unrecognised in his time, mainly because Clement IV died soon after receiving Bacon's large thesis on natural philosophy and the calendar.

Thomas Aquinas (1225–74) rescued the Church from the embarrassment of its own calendar, by arguing that the existence of man and all that man can do is proof of God's existence, so that whatever man can do must be a gift from God. This opened up more opportunities for the investigation of scientific evidence.

THE PLAGUE AND THE PROTESTANTS

IN 1345, Pope Clement VI had gathered all the research necessary and was about to issue a Papal Bull making the required changes to the calendar. Suddenly, a wave of plagues struck, of which the Black Death was the worst, annihilating 30 million people, one third of the population of Europe. Suddenly calendar reform once again seemed less than important.

Another failed attempt at calendar reform had its origins in the increase of trade that had led to the rapid increase in wealth of those at the top of the social hierarchy, including bishops whose opulent garb and surroundings were in sharp contrast to their teachings of purity of spirit and poverty. This, together with the growth of an independent-thinking mercantile class and the introduction of mechanical clocks, began to leave the Catholic Church very exposed in the areas of authority over time.

The Reformation became the next major barrier to changing the calendar, because Martin Luther's objections in 1517 to papal authority, priestly corruption and hypocrisy, and the selling of indulgences, resulted in a half of all European Christians deserting the Catholic faith within the next 50 years.

The Black Death in London, 1349.

The Gregorian Calendar

From 1540, two major developments happened. The calendar reforms of Pope Gregory gave us the calendar we have today and the ultimate acceptance of the heliocentric system provided the victory of science over the Catholic Church.

POLISH ASTRONOMER and monk Nicolaus Copernicus (1473–1543) used his knowledge of recent mathematical advances to predict the motion of the planets more accurately, concluding that the earth and the planets must revolve around the sun. He was reluctant to publish his findings for fear of the criticism that his theories might generate from astronomers and from the Catholic Church. It was, however, over 70 years

after his death that the Church first condemned his work. For centuries though, this heliocentric system had been known to scholars in the East and to some in Europe, but the Church based its worship on the fact of God creating the

Ptolemy's second-century explanation of the heliocentric movement of the sun around the earth. Astronomical observations did seem to confirm this system until Copernicus revealed its fundamental weakness. (From a seventeenth-century plan.)

The Copernican system showed that by placing the sun at the centre instead of the earth, the motion of the planets could be more accurately explained.

earth and therefore explicitly making our planet the centre of the universe. This was supported by Christian philosophers who adopted ancient Greek Aristotelian and Ptolomeic concepts of the sun and the universe revolving around the supremely important earth (the geocentric system). Any threat to this had been deemed as heretical. Galileo Galilei (1564–1642), the celebrated Italian astronomer and mathematician, initially supported Copernicus's work, but was forced to recant in 1633 in one of the last great efforts of the Grand Inquisition to halt the march of progress.

THE TRIUMPH OF SCIENCE

AFTER THE discoveries of Copernicus, the German astronomer Johannes Kepler (1571–1630), with his important work on the elliptical planetary orbits, and Galileo's telescopic observations generated much work on the motion of planets through to Isaac Newton (1642–1727), who established the fundamental concept of gravity and formed the basis of modern science.

Copernicus's calculations in his *De Revolutionibus* showed how far the West had finally assimilated the knowledge and scholarship from the East. He showed the solar

Galileo Galilei.

year to be 365.2425 days long, only 35 seconds out by our current reckoning. Europe had finally taken up the challenge of science it had left behind at the fall of Rome, having lost the vigour and depth of learning inherited from the ancient Egyptians, Greeks, Sumerians and Indians.

POPE GREGORY

THE CHURCH, in the form of Pope Gregory XIII (1502–85) had to accept the gathering tide of evidence, including the astonishing work of Roger Bacon some three centuries earlier, which was undermining its authority by showing that the calendar was simply wrong. Since the *c.* 1450s, calendars had been distributed to anyone who needed to use them, using the new technology of printing, putting further pressure on the Church to be accurate. Pope Gregory set up a calendar commission, consulted widely with church and state authorities and eventually published his Papal Bull of 1582 which transformed the calendar.

Pope Gregory XIII.

THE GREGORIAN REFORMS

- New Year's Day was established as 1 January.
- Ten days were to be lost, 5–14 October.
- Leap days were to be inserted after 28 February every four years, except those divisible by 100, but not 400 (i.e. 1900 was not a leap year, but 2000 is).
- Easter dates were recalculated against a complicated formula which compensated for the differences between the solar, sidereal and lunar years, the different observations of which had originally prompted the gathering call for change.

Acceptance and Denial

It took nearly 400 years for the Gregorian calendar to be accepted throughout the Christian world, and the story of its acceptance reflects the continuing conflicts of faith within the Church and the political imbroglios of its secular partners.

MOST CATHOLIC countries adopted the reforms within the first year, although a further order had to be issued suspending days in February. In the first wave were France, Italy, Spain, Portugal, Poland and Luxembourg, followed by

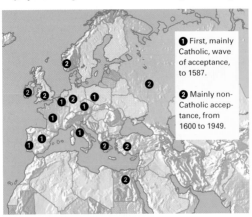

❶ First, mainly Catholic, wave of acceptance, to 1587.

❷ Mainly non-Catholic acceptance, from 1600 to 1949.

Sweden, Bavaria and Austria. The Netherlands, Belgium and Catholic Germany agreed in 1584, Hungary in 1587. Other countries followed at a very different pace: United Kingdom 1752, Japan 1873, Egypt 1875, Eastern Europe 1912–19, including Russia in 1918. China adopted it in 1912, but it took the overthrow of the Nationalist government by Mao Zedong in 1949 to enforce countrywide change.

REJECTION

IN THE first years, while Catholic states accepted the changes the Protestant countries, such as Britain and much of Germany, took much longer to agree to the need for change. This was because the Catholic pope was seen as trying to exert his authority over the whole Christian community after the splits of the Reformation.

This prolonged process of adoption and rejection led, in England, to two centuries of parallel calendar notations, with letters and documents recording 'Old Style' (OS) and 'New Style' (NS).

Martin Luther.

New Year's Day was also erratically enforced throughout the Christian world with 1 March, 25 March and 25 December also being used. This all changed, though, by the time the last major European power, Britain, finally joined the Gregorian scheme in September 1752, bringing, significantly, its colonies, especially the increasingly influential North America.

THE ENGLISH MOB

VOLTAIRE'S FAMOUS jibe, that 'the English mob preferred their calendar to disagree with the sun, than agree with the Pope' was entirely accurate. Although the English court, since the brief reign of the Catholic Mary, had been a cauldron of murderous prejudice both for and against Catholics, Queen Elizabeth I had been consulted during the reform process and all her senior astronomical and scientific advisers agreed with the calculations and the need to change. The newly formed Church of England however, did not, particularly because the most recent Papal Bull had excommunicated Queen Elizabeth for establishing the new Church in place of the papal, Catholic authority. The position was not helped by the attempted Spanish invasion of 1588 which had the blessing of the Catholic Church.

Eventually, in 1750, the Calendar (New Style) Act was passed, invoking Roger Bacon's influence on the reforms, instructing the loss of 3–13 September 1752, one day more than the original reform because the errors of the Julian calendar had pushed the date of Easter back further by this amount.

Queen Elizabeth I.

It is difficult to imagine the strength of feeling, but there were riots in Bristol with people shouting for the return of their 11 days. There were quiet revolts too, of the bankers in the City of London: the changes meant that they would have had to pay their taxes 11 days earlier than the normal year end, which for centuries from the early Saxons had been 25 March (Lady Day or the Feast of the Annunciation, the date on which Jesus Christ was conceived and hence an appropriate start to the year). Prudently, they refused to lose the 11 days of interest on their money, paying on 5 April, creating the precedent which still gives us this date as the end of the financial year, as determined by the Bank of England.

THE EASTERN ORTHODOX CHURCH

SINCE THE excommunications and consequential divisions of the eleventh century, co-operation between the two main limbs of Christianity – the Catholic and Eastern orthodoxy – was at best grimly courteous. At the time of Gregory's papacy, the Eastern Orthodox Church remained furious with the Catholic Church over its lack of support during the fall of Byzantium to the Muslim Turks in 1453 (the Turks added a final insult by changing the name of Byzantium to Istanbul in 1457). Eventually, in 1923, the Eastern orthodox faith adopted the majority of the reforms, but its calculation of leap years for centuries is based on a different system (bringing it slightly closer to the true solar year than the Gregorian) and their Easter dates retain the Julian calendar methods, with the result that there can be a five-week difference in the Easter dates of the Catholic and Orthodox Christian countries.

The sack of Byzantium marked a period of cultural and military expansion by the Ottoman Empire, reaching its zenith c. 1520–50 under Suleiman the Magnificent.

A Summary of Calendars

Calendars have been created and reformed by civilisations such as the ancient Chinese, individual events like the French Revolution and religions like Islam. All such calendars are based on the observations of natural year cycles.

ANCIENT EGYPT

A 365-YEAR day was used from at least 4000 BC, divided into 12 months of 30 days, plus five extra days. Three groups of four months followed the flooding, growing and harvesting

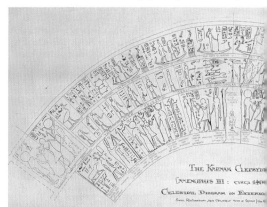

THE KARNAK CLEPSYDRA
[AMENOPHIS III : CIRCA 14..]
CELESTIAL DIAGRAM ON EXTERIOR
Small Restorations from Oblique Time at Sekhet [Jan 6..]

seasons of the Nile. As there were no leap years, New Year's Day moved throughout the seasons over a 1460-year cycle. The adoption of the Julian calendar in 26 BC saw some slight variations, including the new leap days added at the end of the Egyptian year: 29 August of the Julian calendar.

ANCIENT GREECE

A NUMBER of different systems were used, all attempting to reconcile the observable differences between the solar and lunar years. Before 1000 BC, a lunar year with erratic intercalation to bring it into line with the solar year was eventually superseded by more systematic *octaeteris*, an eight-year cycle with five years of 12 months followed by three years of 13 months. In the fifth century BC, the Athenian Meton discovered the cycle to which his name has been lent: the metonic cycle, a 19-year solar cycle of 235 lunar months. This cycle had been independently utilised by a number of cultures.

A celestial diagram copied from the embossed hieroglyphics on the side of a 1400 BC Egyptian waterclock (see page 65).

CHINA

THE ANCIENT Chinese, from at least as early as 2000 BC, used a combination of lunar and solar years, employing leap months to resolve the difference and, like a number of other cultures including the ancient Greeks, leading to independent adoption of what we now term the metonic cycle. Numbers were not used to mark the years, but 22 symbols of nature and cosmology (namely the Chinese zodiac) were combined to create 60-year periods, named after the first ruling emperor of the period. Modern China combines lunar and solar cycles to create a year of 24 two-week periods.

FRENCH REVOLUTION

LASTING FROM 1793 to 1806, this calendar was divided into 12 months of 30 days. A 10-day period was called a decade and days were divided into units of 10 hours. New Year's Day started on the autumn equinox.

GREGORIAN CALENDAR

A REFORM of the Julian calendar, this is based on a more accurate measurement of the solar year, 365.2425 days, and caused the removal of 10 days from the year of 1582 when it was initially introduced. It is now the *de facto* standard throughout the Western and diplomatic world.

INDIA

ANCIENT INDIA, with a cultural inheritance that included Babylonian, Sumerian, Vedic, Hindu and Buddhist influences could boast of almost every combination of lunar and solar calendars. Modern India uses the Gregorian calendar but New Year's Day and the year count differ: the reformed Indian calendar was introduced on 22 March 1957 of the Gregorian calendar, corresponding to New Year's Day of the historical Saka calendar, 1879.

The arrest of Louis XVI at Varennes in 1791. The French revolutionary committee soon adopted its own calendar.

ISLAMIC CALENDAR

THIS LUNAR calendar started in AD 622 according to the Gregorian calendar, marking the date of Muhammad's emigration to Medina. Known as the *Hidjra*, this event gives rise to acronym AH (*After Hegirae*) used after the date. The year is 354 days long with 12 months alternating between 29 and 30 days. Leap years occur 11 times in a cycle of 30 years

and New Year's Day, like the calendar of the ancient Egyptians, moves through the seasons during this cycle.

JEWISH CALENDAR

THE MODERN calendar was defined between the fourth and tenth centuries AD, with dates reaching back to 3761 BC, the beginning of the world as established by Jewish scholars of the period. The lunar calendar has a complicated intercalation which is designed to avoid leap days and months falling on feast days, Fridays and Saturdays. This gives ordinary years of 353, 354 and 355 days and leap years with 383, 384 and 385 days. New Year occurs during September.

JULIAN CALENDAR

BASED ON the solar year this lasts 365.25 days. The quarter of a day was accumulated over three years and added in the fourth. However, the true year was slightly shorter, which meant that the calendar was 11 minutes and 14 seconds too long. This inaccuracy lasted until 1581, by which time European mathematics could compute fractions smaller than a quarter and therefore make more precise predictions.

SOUTH AND CENTRAL AMERICA

AZTEC AND Mayan cultures were amongst the many who used two concurrent calendars, a ritual 260-day calendar of feasts and days of worship and a 360-day year with 18 months of 20 days each, plus five special days (see p. 22).

Incas lament the vanishing sun (due to a solar eclipse), a phenomenon which they saw as a sign of foreboding and death.

A SHORT HISTORY OF CLOCKS

Introduction to Clocks

ANY CLOCK must have a steady repetitive action and be able to display the units of its progress. The later history of clocks is characterised by the determined search of astronomers and mathematicians to measure these units more accurately.

In Europe sundials, waterclocks and the first mechanical clocks only normally gave hours as units until the pendulum movements in the mid-seventeenth century brought accuracy to within about a minute a day. Seconds dials appeared from the 1670s, and stop-watches showing hundredths of a second appeared by 1800. The quartz and atomic clocks of the mid- and late-twentieth century have achieved an astonishing degree of

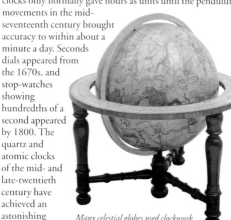

Many celestial globes used clockwork mechanisms to aid their accuracy.

accuracy, so much so that the measurement of time has been wrested from its roots in the natural rhythms of the day, the passage of the seasons.

TIME AND EMPIRE

CLOCK TIME – our time – is now controlled by the vibrations of the caesium atom, so that the second, which used to be expressed as a fraction of the solar year, is now defined by observing the behaviour of caesium.

In this way time has been wrenched from the observations of astronomers, whose work had made them the keepers of time from the ancient worlds of the Babylonians, Greeks and Chinese until the invention of quartz oscillations in the 1930s. Now in the hands of mathematicians and physicists and their calculations of quantum mechanics, time, as relayed by clocks, is controlled by world standards that exceed the accuracy of the planetary motions which astronomers observed so faithfully. Humankind now exercises a collective control over its measurement of time, going far beyond the passive and subjective observations of former eras.

The pendulum clock started the process, which gave its European discoverers an enormous advantage over other cultures in the following three centuries. It marked a level of collective cultural discipline not unlike that of Caesar's Julian calendar, which provided him with a means of defining events throughout his Roman Empire. The Dutch, Spanish, Portuguese and British created empires and could maintain control by using the disciplines of reporting and acting on certain days and at certain times. In this way small, but highly organised nations of Europe were able to conquer and rule more ancient civilisations in India and the Middle East.

Sun Clocks

The use of the sun to approximate the time is as old as recorded time itself and there are many different methods of using the shadow created by the sun.

THE FIRST measure of the hours was a simple stick. The Babylonians, in the plains surrounding the Euphrates river before 4000 BC, used such a stick, which they later developed by dividing the day into 12 daylight hours, marked by the stick's shadow inside a scalloped stone, and 12 night-time hours.

Tibetan priests and Indian fakirs used sundial sticks with a hole at the top and a small peg inserted which cast a small shadow on to a series of units marked on the stick itself.

The earliest known record of a sun clock is in Egypt, during the time of Thutmosis III, sometime between 1501 and 1448 BC: a 30-cm (11-in) simple stone device, in the form of a T which, when pointed at the sun, threw a

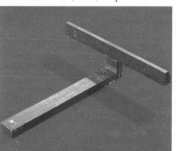

The shadow clock was placed in an east-west position, so the morning sun threw a shadow on the up-turned end of the base. At noon the device was reversed.

shadow down the length of its stem, marking time against lines scored into the stone. It could be used up until noon when it had to be turned over.

Egypt also made extensive use of obelisks, the oldest being built in 3500 BC, marking noon, the longest and shortest days and eventually further sub-divisions. Inevitably the Romans, having conquered most of the known world by the last century before Christ, also used obelisks, such as the one built by Augustus in 27 BC, which still looks out of the Campus Martius in Rome. The ancient Greeks used the measurement of footsteps along the shadow of a column.

Diagram of an obelisk, similar to the one in the Campus Martius in Rome. Obelisks work on the same principles as the sundial, showing two spans of 12 hours, on which the shadow of the sun falls to indicate the hour.

SUNDIALS

THESE CAN be found in almost every society at some point in its development, latterly as ornaments. They were, and can still be, found in a number of different forms; indeed in 30 BC Vitruvius noted at least 13 different types in use throughout Italy, Greece and Asia Minor.

There is an early reference to a sundial in the Bible, *Kings* II, XX:11:

And Isaiah the prophet cried out to the LORD who brought the shadow ten degrees backward by which it had gone down on the dial of Ahaz.

King Ahaz ruled Judea between 740 and 728 BC and these lines seem to refer to the lengthening of the life of the king, using the sort of large stone dial characteristic of the period as a metaphor.

Simple sundials which could show midday and the general passage of the day (morning-tide, noontide, eventide) were placed, hanging down vertically, above doors in medieval Europe, and in the eleventh century, pocket and hand dials were in use, these hand dials performed the function of a pocket watch and were used by farmers and travellers.

Some cathedrals featured a dial on the floor of the nave, with an aperture in the roof so that a sunbeam would illuminate the hours of the day; others featured external sundials.

In the thirteenth century, the Moroccan mathematician Abdul Hussan designed what we would understand to be the familiar sundial, with the *gnomon* (the hand of the dial)

A copy of a tenth-century Saxon pocket sundial found in Canterbury Cathedral. In the absence of such simple devices, a twig placed in the hand served its general purpose.

pointing to the North Pole and the dial itself divided into equal units. Using trigonometry which would have puzzled his European contemporaries, Hussan worked out the formula for making accurate measurements at every latitude.

Since the Middle Ages, the most popular dials were horizontal and became increasingly secondary in use as they provided time for setting mechanical and spring-wound clocks.

Sundials suffered from the weather, though. One intriguing way round this was achieved in first century BC when the Syrians made a gift of a monumental, octagonal waterclock to the city of Athens. The water device ran independently of the eight sundials which were placed on each side of the clock. Each sundial was held by a wind spirit, lending it the name of the Tower of the Winds, and it became the official timekeeper of Athens.

Lincoln Cathedral, with two types of sundial set in one of its external buttresses.

Waterclocks

Waterclocks represented a step away from reliance on celestial bodies and are still in use today.

WATERCLOCKS classically have a hole at the bottom or near the base of a vessel which starts the day full, ending it empty. As the Egyptians discovered, to be successful the angle of the side of the bowl needs to be 70° to ensure an even flow. The earliest known version was found in the tomb of the pharaoh Amenhotep, from 1500 BC, although waterclocks have also been discovered throughout the Middle East, in India and through Europe to the ancient Britons and Picts in Ireland. Sinking bowls, a reverse form of the waterclock, where water flows into the bowl, can still be found in North Africa.

Called *clepsydras* (or 'water thief') in Ancient Greece, small waterclocks were used to limit the time a lawyer could present

his case in court (with plenty of opportunity for bribery by either side), and in the theatre they showed the duration of the performance. Archimedes, a third-century BC Greek mathematician, made a carefully crafted waterclock for his friend Hiero, the king of Syracuse,

Small waterclocks like this were used to time the lengthy summaries of lawyers in court.

while conducting experiments which led to the measurement of an object's weight by displacing water, and the water screw, a method of moving water from one level to another.

The Greeks and Romans took great pride in their waterclocks, making elaborate, ornamented versions and endlessly refining the timing of the flow, trying to regulate the pressure of the water. Bells, gongs, doors with figures and astrological models were added in flamboyant displays of craftsmanship and scholarship.

Cast of an early Egyptian waterclock, dating from 1415–1380 BC, found at Karnak, Upper Egypt, in 1904.

WATERCLOCKS IN THE MODERN WORLD

IN THE ninth century AD, the Islamic world had inherited much of the Mesopotamian, Greek and Egyptian learning, and continued in the tradition of making elaborate waterclocks, adding automata animals and birds which sang on the hour. In AD 809 the Caliph of Baghdad sent such a clock to

Su Sung's 1088 astronomical waterclock has inpsired many attempts to reveal its accuracy including the pen-and-ink drawing (above) and the model (left). They both show its enormous scale and intricasy. The model particularly shows the mechanised Armillary Sphere and celestial globe.

Charlemagne (see p. 35). The clock contained a dazzling display of devices, including a different door which opened for each hour: the appropriate number of copper balls fell out and struck a metal cylinder, marking the hour. At noon automata knights emerged to shut each door.

In China, clock making developed rapidly from AD 200 to 1300: a manuscript by Su Sung describes a massive water tower of AD 1088, 9.1 m (30 ft) tall, with gongs, five doors, mannequins holding bells, other mannequins holding tablets up showing the hour, a celestial globe in the centre and the use of a device invented in AD 725 by Chang Sui, which provided a water-driven escapement. The Chinese invented the escapement device 900 years before it appeared in Europe: its purpose was to regulate the energy created by the flow of water.

Waterclocks, like sundials, had a built-in weakness: they could freeze in the cold. Alphonso X of Castile in the thirteenth century, used expensive mercury instead of water in his *clepsydra*.

The time reckoner from a waterclock, dated c.1700.

Other Simple Clocks

Human ingenuity is evidenced by the many different methods used to measure time, before the development of springs and pendulums.

IN THE Dark Ages, many of the monasteries isolated from the scholarship and learning of the Middle East and the Orient practised some very basic methods of timekeeping. Some monasteries used what amounts to 'monk time'. During the night, where no sundial would work or waterclock could be seen within the abbey or cloisters, a monk would read a set number of pages from the Bible then run to the belfry to ring the bell.

WASTAGE DEVICES

A NUMBER of wastage devices were also employed: Alfred the Great, while he was hiding from his tormentors, promised that if he regained the throne he would spend one-third of every day

Chinese fire clock, which works on the same principles as the incense clock.

in prayer. Once in power he used two candles, 30 cm (12 in) tall, which burned for four hours. He arranged for the horn of boars and deer to be honed down to a transparent thickness so that the candles be protected by the Dark Age equivalent of a lampshade. In this way the candles could be used in any weather.

In ancient China, incense clocks were used, in which a small box containing incense was lit: the lighted incense burned through a maze in the upper part of the box until the end of the day.

HOUR GLASSES

ALTHOUGH THEY seem to have been a relatively late development, it is possible the hour glass was used by Roman soldiers timing their night watches. Contrary to their name, they have been used for timing anything from a minute to several hours. They underwent a flowering of interest in the time of Queen Elizabeth I, when they were often used in debates and for jousts, and by seafarers on their trips to discover the extremities of the known world. They are still used frequently today for the task of timing boiling eggs.

Hour glasses at Easthope church, Shropshire, have been used to time sermons since the eleventh century; this is a Jacobean version.

Mechanical Clocks

The emergence of mechanical devices as a means of
telling the time was the start of a revolution in
timekeeping, relieving the general public from reliance
on freezing waterclocks or cloud-obstructed sundials.

EUROPE EMERGED with some relief from the Dark Ages,
with stability and order spread throughout a landmass
which had suffered barbarism and ignorance to such an extent
that it had lost almost all of its technological advances.

However, a still small voice of consistency had remained
in the cool corridors of the monasteries and abbeys of the
Christian Churches which, although surrounded by rural
populations who used the dawn, noon and twilight as their
markers of time, quietly continued their liturgical labours,
marking hours with bells and calls.

THE FIRST MECHANICAL CLOCK

AS WE have seen, Europe's emergence from the Dark Ages was
partly a result of the influx of ideas and knowledge from the
East, reconnecting Europe with the noble traditions of the
Greeks, Romans and the Babylonians. Some have claimed that
Pope Sylvester II invented the first mechanical clock in AD 996,
having studied mechanical devices in the Islamic court of
Spain. Throughout the Middle Ages, there was only one name
for a clock – horologium – and it is impossible to tell whether

The oldest existing clock in England, made on or before 1386,
housed in Salisbury Cathedral.

water, sun or mechanical clocks are referred to during the twelfth and thirteenth centuries. It is not known precisely who did invent this new phenomenon, but from the early 1300s there is evidence from the hundreds of repair, maintenance and

purchase bills which relate to the use and the cost of mechanical clocks in towers, monasteries and churches.

In *Paradiso*, written around 1320, Dante uses a clock as a powerful metaphor for life, clearly expecting it to be instantly understood:

> *And the wheels in the clock works, which*
> *Turn, so that the first to the beholder*
> *Seems still, and the last, to fly.*

THE CLOCK MOVEMENT

FOR 300 years, these early clocks worked on a similar principle to the reverse action of a water well where a weight is tied to a rope wound round a revolving drum in order to turn the handle. For clocks however, an escapement mechanism (see p. 67) would be needed to regulate the flow of this falling momentum.

The first mechanical movements, in the spirit of the calling and campanology of religious time, provided a prompt for a bell to be struck by human hand. The mechanism was quickly made to strike the bell itself, and eventually

Escapements: Verge and Foliot

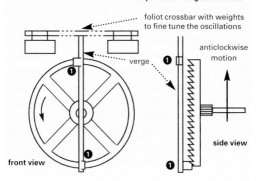

foliot crossbar with weights to fine tune the oscillations

verge

anticlockwise motion

side view

front view

The foliot is fixed to the verge which, using the alternating action of the pallets ❶ regulates the turn of the wheel.

turn hands which could roughly display the hour. It is no accident that our word 'clock' is derived from the ancient Celtic word, *clocc*, from which are derived the French *cloche*, German *glocke* and Scandinavian *klokk*, all of which mean a bell. Earliest known clocks in England include Westminster in 1288 and Canterbury in 1292. The clock of Norwich Cathedral in 1325, built by Roger Stoke, used a simple escapement form called the verge and foliot.

The 6-ft wide Wells Cathedral clock, built in 1392 for Glastonbury Abbey, but moved to Wells by Henry VIII.

SOCIAL CONSEQUENCES OF CLOCKS

THESE DEVELOPMENTS, although only as accurate as many of the waterclocks, did prove to be a more practical alternative and the introduction of the mechanical clock gave rise to a kind of civic time, with some cities constructing a large central clock, emphasising the gathering power of visible time as a means of controlling the way of life beyond the command of the Church.

For the aristocracy and the emergent, expanding mercantile classes this was good news, but for the workers it gave their taskmasters more precise control over their movements. In Amiens, France, for instance, in 1335, the king permitted the mayor 'an ordinance concerning the time when the workers ... should go each morning to work, when they should eat, and when return to work after eating ... they might ring a bell ... in the belfry'.

Clocks also shadowed the rise of numeracy, developing precisely at a time, as we have seen, that knowledge from the East cross-fertilised with European culture, bringing within it the arithmetic learning of the ancient Hindus, then widespread in the Middle East. Clocks gave everyone a more precise sense of units of time. This coincided with the rapid increase of trade from 1100 and meant that traders had to be able to count their goods and make contracts with merchants from the Middle East.

An interior detail (left) and mercury dial (below) of a reconstruction of the first astronomical clock, built in 1364 by the Italian Giovanni Dondi. It shows mean time and was the first to show star (sidereal) time with the motions of sun, moon and the five then-known planets.

Spring Clocks and Watches

Developments from mechanical to finger-ring
mounted clocks took only 200 years of refinement
and craftsmanship.

ECHANICAL CLOCKS were large and heavy, made of
brass or iron and very expensive to build. As interest in
them increased and as technical skills developed, it became
possible to make smaller versions of clocks driven by a spring
instead of a weight which could be taken from one room to
another. The use of the spring-driven mechanism instead of
weights started in the mid 1400s; one of the first known was the
now disputed Burgundy clock of Philip the Good, in 1430,
which has the first recorded use of the fusee wheel to regulate
the diminishing force of the steadily unwinding spring.

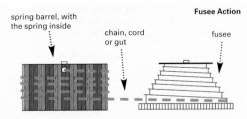

spring barrel, with
the spring inside

chain, cord
or gut

Fusee Action

fusee

*The force of the spring is regulated by the fusee. When the spring is tightly
wound the chain pulls at the top of the fusee, where there is least leverage,
but as it winds down so the chain moves down to the circumference of the
base of the fusee and the weakening spring pulls on a larger diameter.*

Such inventions enabled much smaller clocks to be made and led increasingly specialised craftsmen to dedicate themselves to miniaturisation using special tools to make the parts. In the early, large mechanical clocks, a slightly rough finishing was not magnified into a significant interruption to the movement; on a smaller clock, of course, this was very different.

MINIATURISATION

THE WHOLE of Europe was fascinated by clocks, centering on several south German cities, such as Ulm and Nuremberg, where *c.* 1500–10 Peter Henlein is credited with making the precurser of the modern watch, a small drum clock (Leonardo da Vinci is also said to have designed one, but left it unmade). The art of the watchmaker was highly skilled, as one who could not only make clocks but make them well enough to be very small.

Miniaturisation led to fashionable extremes, with Francis I of France buying two examples of a tiny clock grafted on to a dagger's head. Queen Elizabeth I had a finger-ring clock which contained an alarm and a small prong which emerged to gently scratch her finger. Mary, Queen of Scots was long thought to have had a skull's head watch, although this was later discovered to have been made well over 100 years after her reign.

The 'Mary, Queen of Scots' Skull Watch.

FANTASTIC CLOCKS

IMPROVEMENTS IN accuracy were made as modifications to the escapement by the Swiss horologist and the greatest clockmaker of his day, Jost Burgi (*c.* 1552–1610) who, in 1580 devised one of the most striking pieces of the era, a mechanical globe clock with a golden astronomical globe as the centrepiece. He was also responsible for introducing a novel and improved escapement, the 'cross-beat'.

Habrecht clockwork globe and stand, showing the inner workings of such pieces.

Other fantastic clocks were made, with automata, astrological maps, astronomical tables and fine arrays of bells.

The cost of such timepieces was inevitably very high. Even the less-expensive portable clocks or watches could only be afforded by wealthy clients. They were still inaccurate and, as such, they tended to be a physical manifestation of the power of their owner, paraded as symbols of wealth and craftsmanship.

An astronomical table clock with dials showing the lunar cycle and dates for both the Julian and Gregorian calendars. It struck on every hour and every quarter.

They were used extensively as ambassadorial gifts in much the same way that Haroun al-Rashid, the Caliph of ninth-century Baghdad, sent his elaborate and stunning clock to Charlemagne 700 years earlier.

The oldest celestial clockwork globe in existence, by Johannes Schoner, 1533.

The Pendulum Clock

Until the middle of the seventeenth century, clocks did not need a minute or seconds hand because they were only broadly accurate to the hour.

IN 1582, Galileo is thought to have studied the pendulum's motion after watching a lamp swing in the Cathedral of Pisa. His designs were not completed however, leaving Dutchman Christiaan Huygens, in 1656, to make the first pendulum clock. Clocks were only accurate to within 15 minutes at best, but the pendulum device brought this down dramatically to approximately 10 seconds, making the minute hand a useful addition. In 1671, William Clement is said by some to have first used the anchor escapement, using a long pendulum and keeping even better time. These more accurate clocks were

then made to go for more than one day, most going for eight days, but some were made to go for a month and some even for a year.

These longer duration clocks, however, needed much heavier weights, leading to the development of the long case (called the grandfather clock when over six feet tall). The case provided a strong support for the clock mechanism and its now heavy weights while keeping its motion protected from interference.

A reconstruction of Huygen's 1656 pendulum clock.

An early example of a pendulum clock, made in Paris in 1680 by Isaac Thuret, under the direction of Huygens. Thuret was clockmaker to Louis XIV.

SCIENCE AND IGNORANCE

THIS QUANTUM leap in accuracy coincided with Newton's great discoveries concerning the movements of the planets of our solar system, publishing his laws of gravity, light and motion. Our long-accepted position at the centre of the universe was fatally undermined by such dramatic seventeenth-century scientific advances.

Paradoxically, this greater accuracy highlighted the compromise which our 24-hour clock represents. The unit of one day displays an average time from the start of one day to the next, because over the period of one year, the precise measurements of the 24-hour cycle depend on the position of the earth on a given point in its rotation around the sun. Because these vary due to the angle of the earth against the sun and the elliptical, rather than circular rotation, the unit of one day

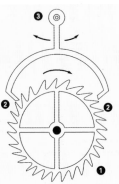

Escapements: dead-beat

The dead-beat is a highly accurate form of anchor escapement providing constant slight pushes to the pendulum.
❶ *escape wheel*
❷ *the escape wheel teeth fall successively on the face of the two pallets which are designed to reduce recoil in the wheel while receiving impulse to keep the pendulum swinging.*
❸ *the pallets connect to the pendulum.*

measured by the clock was, and still is, slightly different to the day as experienced by the observer of the sun.

The march of science was characteristically scorned by many of those who still relied on the natural cycles to determine their lives: in a village near Chester, a local memorial for the clockmaker Peter Clare was suitably mocking:

> *Here's the cottage of Peter that cunning old fox,*
> *Who kept the sun right by the time of his clocks.*

In spite of such widespread attitudes the believers in science persevered and in 1715 George Graham increased the accuracy to under a second per day, by improving the dead-beat escapement (an invention made in the 1670s). This was one of a number of developments which were designed to reduce the effects of friction and slight irregularities in the motion of the train in the clock mechanism.

Developments in accuracy produced timepieces such as these two in the 1780s. Left: a pocket watch by John Arnold and Son, based on the same principles as Arnold's marine chronometers. Right: a piece by Jean Antoine Lepine of Paris who replaced the continuous top plate of the watch movement by a series of bars which made the watch very thin.

The Race for Accuracy

The last 150 years have seen the rapid acceleration of technological advance mirrored by the production of ever more accurate clocks.

MASTER AND SLAVE CLOCKS

THE DISCOVERY of electricity was inevitably harnessed by clockmakers. The electric master and slave clocks in the 1850s, such as the one adopted by the Royal Observatory at Greenwich in 1852, used electric contact points to provide steady, regular beats which could drive not only the movements of their own dials but, through cabling, a number of slave dials at

other locations. In 1889 Siegmund Riefler invented an almost completely free-swinging pendulum which, because it had less friction from the escapement to counter, is said to have achieved accuracy to one hundredth of a second per day and became the astronomical standard. A decade later, in 1898,

The Shepherd master clock with two slaves, used at Greenwich from 1852.

the first true pendulum-free clock mechanism was demonstrated by R. J. Rudd, eventually to be replaced by the Shortt clock (invented by W. H. Shortt) which became the choice of all the main observatories in the western world from 1921 onwards. The master pendulum, isolated in a partial vacuum, was entirely independent of its clock mechanism, enabling it to concentrate on maintaining strict accuracy undisturbed by mechanical tasks like moving the hands around a dial. Accuracy was increased to within one second per year.

THE QUARTZ CLOCK

A MAJOR advance was the use of quartz for timekeeping. Quartz clocks and watches had the advantage of ease of manufacture, with no gears or escapements to be set up or to disturb the oscillations. The principle of the clock was based on the piezoelectric property of quartz crystals: placed in a circuit, in a clock, the combination of the mechanical stress and electric field causes the quartz to vibrate and generate a constant

A quartz clock such as the one at the Greenwich Observatory, used from 1942 to measure to one-thousandth of a second.

signal which can turn hands on a dial or create an electronic display. The first quartz clock was developed in the USA, with the Greenwich Observatory installing one in 1939 which has attained accuracy to within two thousandths of a second in a day.

THE CAESIUM CLOCK

THE SEARCH for accuracy has reached its ultimate, current goal with the institutionalisation of atomic motion. All chemical elements possess a resonant frequency, an ability to emit and absorb electromagnetic radiation which is stable under almost any range of temperature, pressure or humidity and so can be used as a reference point for time on all parts of the globe and, indeed, on satellites above the earth and spaceships travelling from it. Developments in high-frequency and radio-wave measurements made it possible to create a device that could generate electromagnetic waves which could then be made to interact with atoms. In 1957 the USA's

The original 1956 caesium resonator which led to the development of the atomic standard of time.

National Institute of Standards and Technology (NIST) created a caesium atomic beam device based on these discoveries, and found that they could achieve the most accurate results in the history of time measurement. Some subsequent measurements have been found to achieve accuracy of one millionth of a second in a year. Such accuracy meant that the second became the preferred international scientific standard unit, measured as 9,192,631,770 oscillations of the caesium atom's frequency.

By 1967 the frequency of caesium atoms was formally recognised internationally as the universal measure of time.

Three caesium clocks and monitoring equipment at the National Physical Laboratory, for the two hydrogen masers which make up the UK time system UTC (NPL).

The Standardisation of Time

The adoption of a single time standard in 1884 brought timekeeping into a new era of simultaneous communication.

THE GREENWICH OBSERVATORY

THE ROYAL Observatory at Greenwich was established in 1675 by King Charles II and from those first years it has remained a world authority on matters of astronomy, navigation and timekeeping. The observatory gathered astronomical data and measurements of the earth's rotation, providing information essential for navigation and communication. Throughout the whole of the eighteenth century, it played a central role in encouraging and testing the developments of accurate measurements for navigation at sea with the chase for the great Longitude Prize (see p. 96).

The time ball at the Royal Observatory drops at 12 noon GMT. Mariners on the boats and docks of the Thames observed this and set their clocks by it.

Sir Home Popham's telegraph system was adopted by the Admiralty in 1816.

The Observatory consolidated its authority in the nineteenth century. As the Industrial Revolution swept through Britain, the construction of the railway from the 1830s represented a major advance in communication and trade: the introduction of railway timetables revealed the potential for chaos as the impact of the 30-minute time difference between Land's End and Lowestoft was realised. In 1880 England, Scotland and Wales agreed to use a single standard of time, governed by master clocks at Greenwich, adopting the Greenwich Mean Time. The new electric telegraph, another major advance of the Industrial Revolution, was used to distribute a time signal from a master pendulum clock to time signals around the country.

The first successful electric telegraph was the 1837 Cooke and Wheatstone's. It was used on the railway and, by 1838, was sending public telegrams.

THE RADIO PIPS AND BEYOND

THE DISTRIBUTION of the time standards was in a perpetual state of refinement so that by the early 1920s and the coming of radio, a great opportunity was seized for a universal audible confirmation of the correct time. The successor to the

original six pips is still being transmitted every hour on BBC Radio 4 and the World Service. In the USA, a similar service was established by the US equivalent of Greenwich, NIST, which also transmitted signals and messages on the telephone. The signals were accurate to within one thousandth of a second when corrected for distance from the receiver.

Advert for Power Tone radios, c. 1925.

By 1975 signals were distributed to the whole of the western hemisphere via satellites above the equator, with two weather satellites broadcasting a time code accurate to within 100 microseconds. The United States Naval Observatory (USNO), established in 1830 as the Depot of Charts and Instruments to care for the US Navy's

Advert for 1920s Radiola.

chronometers and charts, is now the *de facto* manager of precision time: a system of caesium clocks and hydrogen maser clocks constructs a time reference point which produces signals to make the USNO master clock. This master clock is used in a time server which can be dialled into on the Internet or can be received by a radio-controlled watch. Both Britain and France have similar time systems and together they form part of an international network which controls our standards of time.

UNIVERSAL TIME

IN 1972, an agreement was reached to manage the difference between the time standard based on the earth's rotational time (GMT) and the new atomic standards based on the oscillations of caesium which were independent of the earth and its astronomical motion. This new standard was called Co-ordinated Universal Time (its acronym, UTC, is a typical political compromise because the French and the English both wanted to use the initials from their own respective language): under this system, caesium time is held until nearly one second difference appears with GMT, then a leap second is inserted. The International Bureau of Weights and Measures in Paris receives data from observatories around the world and co-ordinates the necessary changes.

This radio-controlled wristwatch can receive signals from within 1000 miles, from its aerial in the leather strap.

TIME ZONES

TIME ZONES became necessary in the USA by the 1860s, because trains could travel hundreds of miles in a day. Until then people relied on local time, which changed one minute for every 12 miles travelled in an east-west direction in mid-America. This gave over 300 local times for a journey across the USA, so 100 train zones were established.

In 1883, the USA was divided into four time zones based on the 75th, 90th, 105th and 120th meridians. At noon on 13 November 1883, the electric telegraph distributed GMT to all the major cities, so that the local authorities could adjust their clocks to their new zone's time.

In 1884 the International Meridian Conference in Washington DC established world zones with 24 standard

Telegraph station overlooking New York Bay (1838 print).

meridians every 15 degrees east and west of 0 at Greenwich, as centres of the zones. The International Dateline, the passage from one day to the next, was drawn on 180 meridian in the Pacific, the frantic zigzag of the international dateline reflecting the desire of islands and nations such as Fiji, Kiribati and Samoa not to be split.

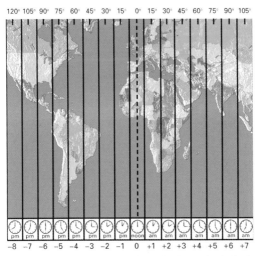

Section of a time-zone map showing the meridian lines and angle of each meridian (top); clocks measuring time before and after GMT at noon, and the number of hours to be added or subtracted to provide GMT. The 24 time zones extend from -12 to +12 GMT.

TIME AND NAVIGATION
The Need for Accurate Time

> By the eighteenth century, the search for finding longitude at sea was one of the greatest scientific inquiries of its time.

EARLY NAVIGATIONAL aids concentrated on maps of the stars, either in the form of celestial globes and table maps, or devices like the astrolabe (which showed the position of the stars at different dates and latitudes).

As the sea-going nations of Europe – England, Holland, France, Spain and Portugal – began to strike out in the fifteenth century, trying to find their way to the New World and the Far East where they could bring back bullion and spices, the cost of such journeys became highly significant. Lack of knowledge of the exact position at sea often led to lost crews, lost ships and huge financial losses. Two coordinates are necessary for accurate position at sea: latitude and longitude. The former had long been found by such devices as astrolabes, but longitude was much harder to find.

The earth is essentially a giant clock, which takes a day (24 hours) to complete the revolution of 360°, so each hour is 1/24 of this,

A nocturnal from 1589; these could be used like clocks at night by aligning the arm to constellations like The Great Bear. The time could be read by feeling the teeth on the edge of the dial.

namely 15°. Longitude can therefore be measured by the time difference between local time (measured by, say, checking the position of the sun), and firm knowledge of a known reference point, like Greenwich Mean Time (GMT).

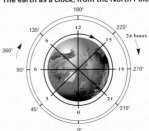

The earth as a clock, from the North Pole

From the late Middle Ages, the search began for a reliable time device which could give the mariner accurate GMT while at sea. Early mechanical clocks were simply too inaccurate without a stable platform on which to run.

THE PRIZE

BY THE end of the seventeenth century, a rash of prizes and awards injected a surge of scientific endeavour into the creation of a reliable solution. King Philip of Spain, for instance, offered 100,000 crowns and the Dutch government 10,000 florins.

In 1714, the British government offered a prize of £20,000 for a clock which could boast an accuracy of one half a degree after a voyage to the West Indies and back.

Arabian astrolabe, 1054 (back view); these gave the mariner a picture of the stars and planets at different times, dates and latitudes, enabling him to measure his position with the naked eye.

Harrison and the H Clocks

JOHN HARRISON was a determined Yorkshireman. A carpenter by trade and self-taught clockmaker, he took a different approach to the few others trying to win the Longitude Prize, some of whom were astronomers. The Board of Longitude had been established to examine claims to the prize and included the top scientific names of the day, including Sir Isaac Newton, until his death.

The mechanism of H1, weighing 75 lbs and originally housed in a four-foot square cabinet.

Over a period of 40 years Harrison made four significant designs as prototypes; his development work and ultimate success were made more difficult by a combination of his own stubborn high standards and insidious competition from those in more powerful positions than himself (at one point, Nevil Maskelyne, a supporter of a rival method, was made Astronomer Royal and took a place on the Board which was judging Harrison's work!) and the understandable caution of those who were responsible for administering the enormous prize money. In 1735 Harrison presented his first attempt, the timekeeper, now known as H1 which proved to be remarkably accurate.

The H2.

TRIAL BY RIVAL

The H3.

FROM THE point of this first success Harrison's life became entangled in an extraordinary struggle against the bureaucratic institutions which governed the progress of science. His ultimate design, H4, an excellent refinement on his original, was tested twice alongside the astronomical methods championed by members of the Board. He did, however, receive half payment of the full prize for the elegant achievement of H4, but was never fully recognised by them as the winner of the 1714 award. In 1772, four years before his death, Harrison's son William – who had continued his father's work – petitioned King George III for help, exasperated by the injustice of the Board's behaviour. He received full support and an eventual award from parliament for the industry and inspiration of his father's life's work.

A copy of H4 had been taken on Captain Cook's second voyage and was used to chart the South Sea Islands. This copy was made by Larcum Kendal who went on to make cheaper versions of H4 in response to the Board's new challenge to find a timepiece which could exceed Harrison's efforts and would be cheap enough for everyday navigational use. John Arnold probably succeeded above all: by 1779 he produced a pocket timekeeper, his no. 36, and with it invented the use of the term 'chronometer', moving on to a kind of 'batch production' with a version of it within several years.

H4 was a large pocket watch, tiny by comparison to the H1.

TIME AND THE WAY WE LIVE
Inner Time and Social Time

I just have a different clock, man.
Keith Richard of the Rolling Stones, 1998

EVERYTHING WE do is defined by our attitude to time. Some people try to control it, some let it control them, some indulge in an orgy of social conformity, finding succour and stability in a received standard of time, while others like Keith Richard revel in a strong sense of personal time which dances to a different rhythm. It is common for artists, writers and musicians to choose to shape their creations during the night. Somehow the stillness of the air and the intense focus of personal activity provides some spark of exhilaration which is stifled during the clattering social strictures of the day.

In families, there are individuals who determine the received view of time – usually a parent – and others who, due to age, circumstance or attitude, do their best to flout it. The teenager who stays in bed all morning is passing though the same physical hours as the small child who gets up at the crack of dawn expecting to play with his or her parents.

Meal times in large families are often very ordered affairs.

MANUFACTURED TIME

THE HOURS and set-piece times like breakfast and tea time are social constructs which provide a basic support structure either to accept or work against. These social times are crucially artificial, manufactured from a series of compromises between lunar, solar and atomic modes of time and a decision by governing bodies to adopt a particular method of timekeeping, conforming to the festival and sacred needs of, in particular, the Christian religion. Inner time, the sense of personal time which connects to the natural rhythms of morning, noon and evening has become less strictly important.

Even so, most people do not rush to telephone the speaking clock to synchronise their watches to the second. We talk of doing something in 'a couple of minutes' meaning very soon, or going out in 'an hour or so'. It is not strictly necessary for us to obey the artificial rules of our clocks, merely to retain a general sense of importance, to intuit the extent of the time and match it to the importance of the task. Zen Buddhist's believe that *prajna* knowledge allows us to rise above the fussy minutiae of rational progressions and to respond in sync with a sense of personal time which puts us in touch with our past and our future.

TIME CONTROLS US

WHEN WE are ill or tired and we sleep for longer than usual, waking up disorientated, one of our first responses is to find out the time, or the day. This immediately reassures us, gives us a reference point for our place in that moment, in that place.

Similarly we are children of our time. Although we suffer the progression of linear time, we grow older and accumulate knowledge; in one sense we are fixed in time at a series of particular events in our lives. War babies were brought up on rations and have a

The school bell and the times for lessons are important social controls.

well-developed sense of community duty; children of the 1960s' baby boom assume a freedom of thought and have a strong sense of personal rights above collective rights. 'I was born in a slum', 'I was born in a mansion', 'I was a slave', 'I was born during the time of Martin Luther King'. The fixed moments of personal time hold us and inform our behaviour for the rest of our lives.

INNER TIME UNDERMINED

TIME CAN be more insidious than simply passively holding us, marking moments of history. Many people roll down the mountain of their lives, gathering momentum, passing exams, gaining qualifications, gaining promotion, impelled always by the deadlines and expectations of peers and authorities. The anxiety of our impending demise, at almost any age, provides a compulsion to finish a project or to reach the end of a journey.

This becomes more apparent when we are obliged to wait. Our personal rhythms often find it hard to deal with the socially sensible system of queuing for a doctor or a film. Somehow it forces us to realise that time operates beyond our control.

A 1920s English mother plays with her baby, knowing that the fears and the aftermath of the Great War were over.

Controlling Time

SOME PEOPLE simply abuse time, by being late for meetings or simply not turning up at all. For some, punctuality, or arriving early in order to be punctual, is an exercise in underachievement – there are so many things which could be done in the minutes or hours spent waiting – just how many 100-metre world records could be broken during a several minute wait for a meeting? Indeed, it is often said that a task is more likely to be done quickly if it is given to an already busy person than one who has very little to do. However, society and its social time conspires against this, because punctuality implies courtesy and conformity to a collective view subscribed to by the person who set up the meeting in the first place.

When they are born, babies have simple needs for shelter, warmth (both emotional and physical) and sustenance; they wake at any time of the day or night to demand fulfilment of one of their needs, unaware of the niceties of social time. Over the years of childhood, these babies are tamed, trained to expect food at set mealtimes and go to bed at respectable hours. The parents and schools conspire to reinforce the social structures which have created the society in which we live.

Some people attempt to control time beyond their life, through cryonics.

As we grow older, we start to try to slow down the grinding inevitability of relentless social time. We have facelifts, we exercise more, we eat more healthily. We attempt to reassert a sense of inner time in rebellion against the impersonal march of time and its conspirators of social time.

A working parent with children is under constant time pressure, both enjoying the time spent with their children and worrying about not getting everything done.

Time and Behaviour

Time has an explicit effect on our behaviour as we react to disruptions and attempt to adhere to our inner time.

TIME AND PERCEPTION

THE MARCH of social time is generally less important to us than how we feel. As we have seen, the daily, often lifelong, interplay between our inner time clock and social time

Time passing at Liverpool Street railway station – too quickly for commuters who rush for trains, too slowly for those waiting for arrivals.

produces a natural pattern of life to which we must conform. However, it is the perception of the variety of circumstances of our experience of our lives which fundamentally informs how we feel and drives our perceptions of the way we live.

TWO CARS

WE ALL travel through a linear, social time at the same rate, measured by our clocks, watches and the pips on the radio, but the differences between us are the same as those between a sleek new Aston Martin DB7, a fine comfortable sports coupé, and a rather battered 1968 Morris 1100 which has travelled over 100,000 miles. On the motorway, both their speed dials tell us that their speed is 70 miles an hour, but if you are the driver of the Aston Martin, the drive is smooth, the engine urging you to break the speed limit as you listen to your CD player. If you are the driver of the Morris, however, the vibration through the worn axle makes your hands judder as you grip the wheel, there is a high, complaining whine from the engine compartment and the old radio is blasting out a mixture of interference from the car's imperfect electrical circuits and some distant, poorly received radio station.

As we pass through life, through time, sometimes we are that Aston Martin,

1874 Punch cartoon: 'I practised half-an-hour, but then that was by the dining room clock. When I practised by the clock on the stairs, it was three quarters of an hour, because that's slow.'

sometimes we are the Morris: for children, long holidays and long journeys can be interminable, whereas for their parents the change from the patterns of a working life are such a welcome relief that the return to them comes all too quickly. When we are feverishly excited by a project, while painting, writing or

Factory work, like potato grading, can make time drag interminably.

making costumes for a local theatre group, we might easily concentrate all night without a sense of passing time, lifting ourselves to another place, suspending our normal sense of time, transported by the fire of an obsession. The anticipation of going to a party, though, can make us feel the drag of the same length of time as we wait for the appointed moment.

There are big differences in the perceptions of the young and the old, those active and those passive, between those in rewarding jobs and those who hate their jobs. It is this time which is most important to us at the personal level.

A peaceful moment, Aragon in Spain.

TIME AND SLEEP

AS ADULTS we generally spend one third of our day asleep. Babies spend two thirds of their time in this state. A donkey spends less than one fifth of its time asleep, while an opossum needs over 19½ hours of sleep a day.

The languid time of love – 'Flirtation' by Soulacroix.

The significance of this is the effect of its disturbance. Physiologically our bodies respond to sleep and the need for it by altering, amongst other things, our hormonal and metabolic states to adjust for a period of inactivity where the ingestion of food and liquids are impossible. These are inevitable daily states whose natural rhythms continue, whether we conform to our needs or not. If we stay up all night, the coming of the fresh morning air still seems to revive us even though we clearly lack the benefits of sleep; the body is still dancing to its own tune as though we had. However, the deprivation of sleep over time leads to a gathering sense of disorientation and lack of alertness which has been harnessed by the torturers of all nations in their wars with their fellow human beings.

JET LAG

JET LAG is a consequence of the massive disturbance of such rhythms. If you travel across longitudes, losing several hours at a time across the time zones, you find that your inner time retains its own sense of what you should be doing. The incidence of chronic tiredness, gastrointestinal disturbance, headaches and malaise are a function of the body trying to force you to conform to its own rhythms instead of changing to accommodate the new day you arrive at in Los Angeles, eight hours to the west of your London departure point.

TIME AND DRUGS

THE USE of drugs in the twentieth century expanded dramatically, from the cynical introduction of opium by European and American traders, supported by their governments, giving them economic control in many parts of China. Modern narcotics such as heroin and recreational drugs like ecstasy introduce chemical and synthetic forces into the natural rhythms of the body, distorting them to such confusion that the symptoms of 'cold turkey' are often an extreme form of the effects of jet lag.

TIME AND MEMORY

MEMORY, THE simple recall of past events, can be severely affected by jet lag, sleep deprivation or drug taking. It can be a release from the worries of the past or it can disorientate you by disconnecting your sense of what you are today from how you arrived at this point. Sufferers of the medical condition Alzheimer's disease can suffer terrible switches of consciousness and blankness which fundamentally undermine their ability to function normally.

Time drifted in the nineteenth-century opium den.

Time in the City, Time in the Country

Lives in towns and cities have always been very different from those in the villages and farms of the country.

24-HOUR LIFESTYLE

TIME, PARTICULARLY in the major cities of developed countries, has become detached from the normal rhythms. We are, palpably, exhausting time and exhausting ourselves by accelerating the means of communication and enjoyment.

All-night clubs and cafés in New York, London or Manchester are no longer unusual. We can shop all night as the major supermarkets compete against one another for that extra percentage of market share. We can ring banks and credit-card companies all night, on

Open all hours in Times Square, New York.

Christmas and New Year's Day. We can send e-mails and news via the Internet to people on the other side of the world, whose time zone gives them daylight hours while we make the telephone cables sing under our moon. The world economy now has to function 24 hours a day: before the London trading floors close, the New York exchange takes over and then the

The Nasdaq stock exchange merged with an all-night club – two modern symbols of fast-moving, frenetic lifestyles.

Chinese and Japanese exchanges operate until London opens again. Music can be downloaded in 'realtime' from the Internet, we do not have to wait to go to a shop and buy it.

RURAL TIME

THE RHYTHMS of life in the countryside are increasingly different from those of the city. More in touch with the natural time of the moon and the seasons, people outside the accelerated whirl of all-night moneymaking are still in touch with an era where clocks were less important than the stars, where observation and instinctive understanding of the world

around them produced the oral
knowledge expressed in the form of
superstitions, old wives' tales, riddles,
songs and nursery rhymes.

*Hedgehog hibernating
from the cold of winter.*

Through the domination of the
Romans in Europe, the Dark and
Middle Ages, the enlightenment of the
Renaissance and the flowering of
scientific learning, to the beginning of
the third millennium, rural communities have used the oral
traditions to convey meaning about their time and show
people how to interpret the world around them.

Forms like the rhyming couplet were easy to remember and
gave important information about events in time:

> *When dew is on the grass,*
> *Rain will never come to pass*
> *When grass is dry at morning light*
> *Look for rain before the night.*
>
> American traditional rhyme

> *Red sky at night, Shepherd's delight,*
> *Red sky in the morning, Shepherd's warning.*
>
> British traditional rhyme

Science can explain that rising dust in the atmosphere turns
the sky red because of refracted light and that its timing heralds
either rain or sunshine depending on the relative warmth in the
air. Traditional rhymes, though, tell us what to expect and how
to interpret the evidence so that if we are skilled or experienced
enough we can predict what will happen tomorrow.

Liturgical Time

Frère Jacques, Frère Jacques,
Dormez vous? Dormez vous?
Sonnez les matines, Sonnez les matines,
Din, dang, dong, din, dang, dong.

French traditional rhyme

S YOU LISTEN to this familiar children's
rhyme, it is interesting to realise that it refers to a
form of time observance which exists beyond
both the natural rhythms of time and the endless
time abuse of the city. Liturgical time, characteristic of
monastic Roman Catholic worship, had its origins in the early
Christian Church in the third century AD. Based around set
hours for prayer, it provided Christians with a way of differen-
tiating their address to God from that of Judaism and a way of
establishing a controlling pattern on the lay Christians, who
relied on religion to provide their social structures.

The Cistercian order in the twelfth century, followed by the
Benedictines, established the order of liturgical time which still
exists today.

DIVISION OF THE DAY

THE LITURGICAL day is divided into Mass and Office,
matins being the first of eight services held throughout the day
and the most difficult for the monks to adhere to, hence the
French rhyme above, because the monk had to arise while it
was still dark in order to toll the bell for matins. This was
followed by lauds at sunrise and four hours of chanting which

occurred on the first, third, sixth and ninth hours of the day.
A period of work ensued for the afternoon, with vespers
carried out in the evening, compline at dusk and a service
at night, mass.

Various forms of Gregorian chant formed the basis of the
worship: the monks in some orders spent most of their waking
day singing and in prayer on a series of strict times, singing the
entire book of psalms in a week. The discipline of liturgical
time, with its days fully independent of the seasons and the
moon, are designed, through its devotional purity and worldly
denial, to absorb the sins of the rest of the population.

Such precision of
dates required a
calendar which could
be relied upon and a
received opinion to
support the faith of
the monks who prayed
in a certain way,
providing a primary
motivation for
the reforms of the
calendar.

*Muslims face towards
the holy city of Mecca to
pray at dawn, noon, mid-
afternoon, sunset and
nightfall. Islamic months
begin at sunset on the day
of visual sighting of the
lunar crescent.*

Time, Power and Institution

Each City has four gates, at each of which there are five trumpets, which are sounded by the Chinese at certain hours of the day and night. There are also in each city ten drums, which they beat at the same time, the better to show their loyalty to the emperor, as well as to give knowledge of the hours of the day and the night; and they are also furnished with sundials and clocks with weights.

Sulaiman al-Tajir, *c.* AD 850

SULAIMAN, a traveller of the ninth century, succinctly captures the many elements of institutional or monumental time: a reckoning of the hour, the public display as a means of controlling the populace, a sense of self importance in the magnitude of the announcement of time, and a significant amount of flattery for the ruler of the city, to show that they are not only in charge, but those conveying their obeisance are only too willing to render it.

Big Ben.

THE PACE OF CHANGE

IN EUROPE, large clock towers started to spring up once the mechanism for the mechanical clock became widely understood. For churches, whose public duties included the ringing of hourly bells, this was a natural extension of their duties. Salisbury and Wells Cathedrals were amongst the first to be built, but by the early 1420s, almost every major city in Europe had a civic clock tower or churches with the new inventions. In 1389,

Rouen Cathedral added a clock which, rather inaccurately, displayed quarters of the hour and was the first device to measure 12-hour units rather than the full 24. In 1370, Charles V of France brought the German Henry de Vic to Paris to make a huge clock on the façade of the Palais de Justice. King Edward III, in 1335, built the first clock tower to stand on the site in Westminster, now occupied by its successor which houses Big Ben. Here we see the association of authority with time and its potential for controlling the populace and as a demonstration of regal power by being able to make such an object.

These clocks were all spectacular, not only because they were a novelty, but because of the elaboration of the designs, many of them including astronomical charts which ran in sync with the time hands, and automata to strike the bells.

Today, monumental clocks still exist and range from special commercial commissions, like the old *Time Life* magazine building in Bond Street, London, to local town halls.

The astronomical clock at Prague Cathedral.

Sport and Speed

Most sports use time as an intrinsic part of
their rules, adding to their fascination for the
participants and spectators.

THE SPECTATOR

THE EXPERIENCE of time varies from one sport to another
and from the perspective of the spectator or the player, the
manager or the steward. Once again, there is a strong sense of
personal time, in which a fan of the winning team, such as of

Manchester United during their historic treble triumph in the European Cup Final in the 1998–99 season, is excited and finds time passing quickly. Their Bayern Munich opponents found time dragging as they held on to their lead until the last few minutes, when their hopes and dreams were destroyed by two stunning opportunistic goals from the Manchester United side.

The spectators of different sports experience a similar sense of differentiation: 40-minute squash matches with their frantic burst of activity are the antithesis of the picnic-and-deckchair mentality of cricket. American sports such as basketball, baseball and American football function in short bursts over long periods, with commercial breaks and comfort breaks on the assumption that their fans have very short attention spans. English sports tend to be more dour with fewer goals and longer periods of play.

Time plays a significant part in tennis too, where the speed of the serve in the men's game has transformed the thrill of play into a familiar succession of cannonballs. Top players survive on the power of their serve to an extent that would have embarrassed former generations of players.

Greg Rusedski, who serves at 148 mph.

Manchester United celebrate their Champions' League 2-1 victory over Bayern Munich.

FAST TIME

IN FORMULA ONE racing, the time element is different again. The drivers qualify, in a fixed-time session on the day

before the race, by completing and being ranked on their speed around their fastest single lap. Teams focus on the single lap because the faster the time, the higher the pole position so the bigger the advantage at the start of the race itself. This focus on a single fast lap is in complete contrast to the actual race, in which time is only

In modern Formula One, a fast pit-stop can determine the winner.

Oxford and Cambridge athletics match at Christchurch Meadows, 1868.

important in the catching up of another driver; the most important challenge is to finish ahead of everyone else, whatever the time.

ACCURACY

IN TRACK athletics, faster times and world records are an essential part of the enjoyment and achievement. Since the late nineteenth century, dramatic improvements have been made in the times achieved by athletes and the accuracy of the measurement of the time. In 1864, the first track event between Oxford and Cambridge used quarter seconds to time the races. By 1880, fifths of a second were possible and became regarded as the best degree of response possible from the human hand in operating a stop watch. During the 1912 Olympics in Stockholm, tenths of a second were used, but although hundredths were available during the 1924 Paris Olympics, they were not officially recognised until the 1970s, when technology finally superseded the partial human eye. Subsequently, for several years, two world records were reported, one for the era up to the use of electronic hundreds and one for hand-held, which was found to be slightly optimistic in its reporting.

Vaqleri Borzov: USSR winner of the 100 metres in the 1972 Olympics.

TIME AND THE ARTS
Fine and Graphic Art

The experience of time is expressed through many
different forms of graphic and fine art.

THE RAPID development of clocks in the fifteenth and
sixteenth centuries led to a fascination with the notions of
time passing. Many of the paintings
of Jan Breughel, for instance,
contain spent objects: empty glasses,
empty masks, finished hour glasses,
burnt-out candles and skulls; men
sleeping in art to convey the overall
sense of *ennui*, either as a supporting
theme for the subject of the painting
or, as in Antonio Pereda's *Le Songe
de la Vie*, becoming the main focus
itself. William Blake's visionary
paintings reveal time as part of the
confining body, and he uses the
defining pair of compasses as a focus
for his *Newton* and *Ancient of Days*.

REPRESENTATIONS OF
THE PASSING OF TIME

THERE HAVE been many
paintings from the end of the
Middle Ages onward which contain

the passage of time within the static frame of a single painting. Sassetta's *Meeting of St Anthony and St Paul* (*c.* 1436) shows St Anthony travelling through the desert in the top left of the painting, to his consultations with a centaur in the middle third right, to his welcome by the hermit St Paul at the bottom. The progression of time is shown through the increase in physical size of the characters as St Anthony approaches his destination, St Paul, and the meandering path of life with its

Willam Blake's 'Newton' (1805), celebrating one of the heroes of his age.

forests of tribulation. Three centuries later, Antoine Watteau's *The Embarkation for Cythere* shows a similar technique, this time depicting the journey of two lovers from a bower in the foreground to a distant boat. Watteau painted a long line of interactions between the two lovers from the bower to the boat, creating a series of film-like stills, a technique used more recently for simple video animations where a large canvas is prepared containing all sequences of actions for the story, and the camera focuses on each painted event before moving to the next.

'The Embarkation for Cythere' (1717), by Antoine Watteau.

Newspaper strip cartoons like the *Daily Mirror*'s long-running *Garth*, comic books with page-long stories like *Dennis the Menace*, or longer forms such as graphic novels with entire *Batman* stories, are obvious demonstrations of the passage of time through graphic art. Some of these, though, play with our expectation of the linear progression of frame following frame by breaking the borders or using occasional 'why' pages for the action to confuse the reader's sense of place and time, as in the *Watchman* series of comics from the late 1980s.

Salvador Dalí's use of liquid clocks also plays with our sense of time, showing this instrument of measurement in unusual positions, such as flowing over rocks, undermining our expectations of linear, endless time.

Scenes from Hogarth's 'Four Times of the Day' (1738): 'Night', and 'Morning in Covent Garden'.

Time and Music

The beat of music, in its independence of the seasons, of the sun and moon, has some similarities to the atomic heartbeat behind all modern clocks.

TIME IS fundamental to music, providing a constant pulse which holds the sound of all the instruments. Here we find time at its most tyrannical, because in order to convey a piece of music in almost all forms, except perhaps the *avante garde* of John Cage's experimental work, in a way that is pleasing to the ear, the instruments and voices must sound together without wandering off on their own.

In all forms of music, the function of time control, or the musical clock, has to be undertaken by someone: a conductor in classical and choral music, a drummer in rock music, or the personal rhythmic playing of the solo flamenco guitarist.

A good conductor, like the late Georg Solti, both interprets the music and keeps the beat for the orchestra.

In studio recording and in the composition of digitally driven dance music by bands such as Prodigy or Underworld, a time code is generated, called SMPTE, which acts as a rhythm for all other sound to be laid on top.

With this album, Prodigy took drum and bass music to the top of the world-wide bestseller lists.

TIME SIGNATURES

THE STEADY pulse normally relates to a particular time signature: 3/4, characteristically used for a waltz such as those written by Strauss or for a ballad by a band such as Boyzone; 4/4, a standard beat for almost every kind of music but particularly of the solid rock of a band like Oasis; and 5/4 which has become a sign of more sophisticated jazz after being used for the first time in 1960 in The Dave Brubeck Quartet's *Take Five*.

STYLES

FLAMENCO RHYTHM is based around the *compas* which is literally the compass, the director of the beat. One such *compas* is used in *solerares*, counted in cycles of 12 beats with the accents falling on the third, sixth, eight, tenth and twelfth beat, which are accented by the stabbing right-hand *rasgueo*. However, the feel of the playing, the *duende*, provides much more drive and passion than the simple accented beat: Paco de Lucia is regarded as one of the greatest flamenco guitarists of all time because he combines explosive passion with a rare technical superiority.

Similarly it is possible to create a piece of perfect music, written on a computer or played by technically brilliant session musicians, but which appears lifeless and dull. For a piece of music to be successful, the musicians and the conductor, if there is one, have to be committed to the mood and soul of

In tribal societies, such as those in Dakr, Senegal, music is often focused on the rhythmic pulse of percussion, which allowed everyone to join in.

the piece, transporting it beyond the simple time patterns on which it is based. Paul Tortellier, the celebrated twentieth-century cellist, was a brilliant performer who managed to wrestle the heart from every performance to create a musical event that reached beyond occasional imperfections of technique.

Charlie Watts, the drummer of the Rolling Stones, is famous for his rock steady, no-frills timekeeping. This style of drumming allows the lead instruments to soar above and around the music in the knowledge that there is a consistent rhythm to return to. John Bonham of Led Zeppelin was a coruscating, flamboyant percussionist who used his drums to compete with the solo indulgences of guitarist Jimmy Page.

Keith Moon was a sensational drummer, adding dramatic flourishes to the songs of The Who, while keeping the simple beat.

Time and Film

The experience of time in films encompasses both the nature of sitting down to watch a movie and the subject matter.

FILMS WHICH deal with the future, for example the *Star Trek* series, or time travel, such as *Back to the Future* and H. G. Wells's *The Time Machine* were all exceptionally popular in their time, teasing out our deep-seated fascination with issues of time. The 1999 remake of *The Mummy* also plumbs our fascination with reincarnation and our fears of the past.

THE LANGUAGE OF FILM

FILMS REQUIRE a satisfying range of the skills employed in fine art, literature and photography. The descriptions, techniques and the language of films give us

obvious direct connections with time: film stills, slow motion, accelerated motion, even the term 'movie' itself – stories told in flashbacks.

John Cleese in 'Clockwise', a film about an obsessively punctual man who suddenly loses control of his life.

Viewers experience a film as a series of images which are designed for them, where time unfolds in front of them like a series of lives and events.

STORYTELLING TECHNIQUE

CLASSIC OLD films used the camera as a witness, a passive observer of the events unfolding before us. We sat in the same room as Humphrey Bogart in *The Maltese Falcon* but our presence had no impact. Modern camerawork takes us into the fray and makes us part of the action, so that our perception of the time and location of what is happening can be affected by our presence. We can be deceived because we are deliberately being shown only a small slice of action so that the full force of dramatic irony – where the audience knows more about a situation than the players on stage or screen – can be given to us or denied to us by the choice of the director.

In the first *Star Wars* trilogy it is as much a shock to us the audience as to Luke Skywalker that Darth Vader, his enemy and persecutor, is in fact his real father. The director, George Lucas, has managed our expectations to the extent that we have lived the life and times of Luke Skywalker though his eyes, at the crucial points where Lucas could have revealed the truth to us but not to his characters.

Christopher Lee as 'The Mummy', in the 1959 film.

Time and Literature

> Literature, by its very nature, passes comment on time and is the greatest explorer of its many different manifestations.

The Moving Finger writes; and, having writ,
Moves on: nor all your Piety nor Wit
Shall lure it back to cancel half a line,
Not all your Tears wash out a Word of it.

THESE LINES from Edward FitzGerald's nineteenth-century translation of *The Rubáiyát of Omar Khayyám*, a philosopher, astronomer and poet at the beginning of the second millennium, are one of the many examples which can be found in literature to use the arrow of time as a device to show that actions once committed, cannot be altered. Most written narrative forms adopt this notion as an expression of the lot of

William Blake evolved a system of beliefs which he expressed through the synthesis of his writing and painting. 'Elohim Creating Adam' shows the creation of man and the imprisonment of the soul in human, time-bound form.

people to submit to their earthly, timebound existence.

William Blake, in his poetry and prophetic works, describes how people are trapped in their bodies and their experience of linear time. In his terms the infinite is to be found within and as an infant, and that to be born is to be bound:

And who shall bind the infinite with an eternal band?
To compass it with swaddling bands? And who shall cherish it
With milk and honey?

'Europe, A Prophesy' (1794)

TIME IN THE STRUCTURES OF LITERATURE

THE EXISTENCE of a novel, collection of poetry or a play in book form challenges our perception of its place in time. It was written by its author at a certain time, performed or read not only at another time, but by several succeeding generations, each of which brings with it a new set of perceptions and interpretations not necessarily intended by the author. *Macbeth* 'was' written by 1606 but 'is' conveying itself at any given moment to an audience somewhere around the world. To these people the play is either successful and pertinent now or it is not worth seeing. George Orwell's prophetic *Nineteen Eighty-Four* has changed in its effect, because we have passed the date of the title, but the potency of his warnings are as strong now as they were in the days after the Second World War when it was penned.

The many different forms of literature address and excite our perceptions of time: Japanese *haiku* can be absorbed in a single breath but prompt many hours of digestion and development in the reader. Novels like Tolstoy's *War and Peace*, one of the longest books ever written, can take several weeks to read. The novels of Charles Dickens were published initially in instalments over several months, with his audience impatiently waiting for each episode, participating in the effect of the cliffhanger event he created at the end of each chapter.

Alice's rabbit, late, for a very important date.

DRAWING IN THE AUDIENCE

THE HISTORICAL dramas of Shakespeare and the experiments in time by J. B. Priestley (such as *Time and the Conways* or *An Inspector Calls*) require the audience to suspend their perceptions of their place in their own time and watch fellow humans act out past lives, to be drawn in to feel the emotions and think alongside the passage of events of the past.

In *Hamlet*, Shakespeare uses a common device of a play within the play so that the audience is watching with the characters, the events of another time. We feel as they feel and our sense of our own time is removed a further step away.

'Alas poor Yorick, I knew him'. Shakespeare explored the phenomenon of time passing, and here makes Hamlet appeal to the past for answers to present questions.

Laurence Sterne's novel *Tristram Shandy* plays with the perceptions of time, shifting his perspective, both bringing us along as a willing observer, then sometimes plunging us into the action so that we might be fooled alongside the characters of the book. It was common for eighteenth- and nineteenth-century writers to address the reader directly to draw them in. The use of letters to narrate a story, in novels like Samuel Richardson's *Clarissa*, enabled a writer to move back and forward in time, to give retrospective perceptions of past events, placing the reader in a privileged position.

TIME AS A THEME

TIME IS often used as a main theme for exploration, Marcel Proust's *A la Recherche du Temps Perdu*, for instance, is a study of transience and *ennui*, and informed a new generation of existential thought. Henrik Ibsen's play *Ghosts* is a classic exposition of the 'sins of the fathers visited on the sons', depicting the present as being trapped by, rather than a progression from, the past.

T. S. Eliot explored themes of the arrow and cycles of time in *The Four Quartets*, which start with the following lines:

Time present and time past
Are both perhaps present in time future
And time future contained in time past.

Samuel Beckett's *Waiting for Godot* explores the mystery/absence of God, eternity and our redemption, with

Estragon and Vladimir finding reassurance in the waiting and the disappointment.

Charles Dickens's *A Christmas Carol*, shows Scrooge being taken back to his past by a spirit guide and to a Christmas Yet to Come. Dickens finally allows him to change his future by a change of heart from his meanness and bad temper.

In Dickens's 'A Christmas Carol', Scrooge is visited by spirits of the past, present and future.

Science Fiction

Time travel, time paradoxes and time slips are all part of literature's purest exploration of time.

SCIENCE FICTION (SF) travels back and forth in time at will. Unlike the realistic writers of the previous pages the SF writer has complete freedom both to write the narrative and create the world in which it appears. Stories explore the complexity of time with a wit and sense of fun usually denied the 'serious' writer, often mirroring the latest newspaper sensations with the most successful containing just enough real science to make their fantasy seem real, but not so much as to restrict the freedom of imagination.

True science fiction started at the end of the nineteenth century with H. G. Wells's *The Time Machine* in 1895 (only a few years before Einstein's astonishing publication of his Special Theory of Relativity), although some have traced its roots back through Mary Shelley's *Frankenstein*, Cyrano de Bergerac's *Journey to the Moon*, Thomas More's *Utopia*, and back to the fantastical myths of Scandinavia and Mesopotamia where powerful beings like Thor or Gilgamesh could alter time with their actions.

Pulp magazines like 'Astounding Stories' and 'Wonder Stories' saw a dramatic rise in circulation in the 1930s and 40s, reflecting the gathering pace of scientific progress with advances like the discovery of quantum mechanics and the expanding universe.

TIME TRAVEL

SCIENCE FICTION is littered with stories playing with the consequences of time travel on the past. Robert Silverberg's *The Assassin* sees a man return to prevent the death of Abraham Lincoln. Michael Moorcock's *Behold the Man* tells us of a man travelling back to save Jesus, but finding himself pretending to be him. A traveller in Dannie Plachta's *The Man from When*, returns to the past with the joyful news that the energy required to send him back destroyed his time, which was, as he explained, only 18 minutes in the future of the past to which he had been sent.

Other stories explore the present consequences: Isaac Asimov's *The End of Eternity* describes a future with an organisation whose task it is to repeatedly return to the past in order to fine tune it so that the future is better. Eventually, everything goes badly wrong.

Captain Kirk and his crew in 'Star Trek' boldly dabble with the edges of all scientific discoveries, like time travel, worm holes and black holes.

PLAYING WITH TIME

OTHER SF stories deal with the complexity of time possibility: John Wyndham's *A Stitch in Time* concerns an old woman who, while thinking about a former lover, is suddenly joined by him as he was then. In *Stormwater Tunnel*, Langdon Jones comes up with the intriguing story of a man who enters a void in time, to answer calls of distress, only to find that it is he who is calling and trapped.

Possibly the most weird of such explorations is Robert Heinlein's *All You Zombies*. A pregnant woman is taken into hospital and during her caesarean operation she is discovered to be a hermaphrodite. The doctors turn her into a man who the narrator of the story sends back in time to meet the woman who becomes pregnant. In a final twist the narrator then pulls the baby, a girl, in to the past and she becomes the woman who became the man who became the baby!

Peter Cushing as the Time Lord Dr Who, with the irrepressible daleks.

TIME AND SPACE

Introduction

To study time at the most fundamental level we need to understand the motion of the stars, the way matter reacts with matter and appreciate how modern science has brought us to the brink of our own origins.

SO FAR we have looked at time at the human level, through our own eyes and ears. Now we need to look at time through the lenses of giant telescopes, satellites and through the theorems of mathematical physics, then out into the universe. There we will see the substance of the matter which surrounds and constitutes us, the forces which act on this matter, and we will encounter the discoveries of modern science which have propelled us with ever-greater acceleration towards and beyond the turn of the millennium.

This chapter explores some of the great scientific theories of the twentieth century: some have only been made possible on the one hand by ever-improving technology such as the study of distant stars which showed that our universe is expanding, while others like Einstein's two great theories of relativity seem to have relied on the sheer creative, non-conformist genius of their creator. We will pick our way through a series of dramatic

The large picture shows the detail of the dust ring around a black hole at the centre of the galaxy called NGC7052 in the small picture. The black hole is probably several million years old and its existence gives us some clues about the beginnings of the universe.

breakthroughs that have enabled us to construct a view of the history of the universe, from the Big Bang through to the present day, with such unnerving accuracy.

The chapter also deals with some of the great questions of time either left unanswered by modern science or left open-ended: can there be a beginning to time? Does infinity exist? In a scientific model of the universe, is there a role for a god?

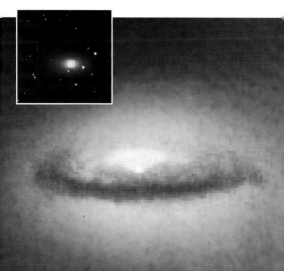

A Short Description of the History of the Universe

The life cycle of our universe can, within some limits, be clearly described. It is an awesome concept to grasp, but we now have sufficient mathematics and theoretical physics to construct a working model of the time from almost at the beginnings of our universe to the present day.

IN THE beginning, from a point of immense (some suggest infinite) density, an all-encompassing explosion propelled every particle of matter away from every other particle. At one-hundredth of a second (the earliest point which can be reliably predicted), the temperature was one hundred thousand million degrees Celsius: so hot that matter, even in its most elemental form, could not form solid bodies.

THE FIRST FEW SECONDS

THE PARTICLES being flung apart included equal numbers of electrons (now essential components of all atoms), positrons (found in some radioactivity), neutrinos (which have no mass or electrical charge) and photons (the quanta of electromagnetic radiation, also containing no mass or electrical charge). The heat of the explosion resulted in property changes to these particles, in which the balance between the components was constantly shifting. The density at this time was still as high as four thousand million times that of water. Other particles included protons (one proton forms the nucleus of a hydrogen

Two star clusters: the yellow, more densely populated areas are estimated to be fifty million years old while the white, more scattered stars are probably only four million years old. Their close proximity suggests that the white stars were created by supernova explosions in the yellow cluster, undergoing the same changes as those stars and galaxies created in the early universe.

atom) and neutrons – which now make up the core of every atom – one each for a thousand million electrons.

At one-tenth of a second the temperature was thirty thousand million degrees Celsius, ten thousand million degrees at one second. As the motion of the explosion continued, the temperature continued to fall as the volume of the universe increased, spreading the energy out more thinly.

At 14 seconds, with the temperature dropping to three thousand million degrees Celsius, more electrons and positrons started to be destroyed than created.

FROM THE FIRST FEW MINUTES

BY THREE minutes the temperature was one thousand million degrees, low enough for matter to start combining in

its simplest form after hydrogen itself: one proton and one neutron, deuterium or heavy hydrogen. Helium also started to form (two protons and two neutrons), both it and hydrogen being fundamental forms within the universe and all structures in it. Most matter consisted of photons, neutrinos and electrons, which were separate from the nuclear matter which was 75 per cent hydrogen and 25 per cent helium.

Finally, after about seven hundred thousand years, the temperature had dropped low enough for electrons to combine with the nuclei to form atoms and eventually clumps of gas which would ultimately create galaxies of stars.

It is important to realise that the first few minutes of the Big Bang explosion created all the necessary conditions of matter and the cooling temperatures needed in order to make stars and ultimately to create us, perhaps even those like us.

TIME AND THE LIFE CYCLE OF A STAR

THE LIFE of an individual star can move in a number of directions. It can end up as a white dwarf or a black hole and along the way it could have a planetary system in which life might form.

Stars are created when large amounts of hydrogen gas coalesce over thousands of years due to the mutual gravitational attraction of the particles. In time, the gas falls in on itself, with the atoms colliding more energetically as this happens. When the collisions are energetic enough, the hydrogen nuclei can fuse together to form helium, causing an explosion of light which we now see from afar as stars shining.

This is how stars like our sun are formed. A telescopic view of the Eagle Nebula (7,000 light years away), showing dense pillars of hydrogen gas and dust with regions of high density which will eventually become stars.

The spiral galaxy, NGC 7742 with a black hole in the centre and a ring of young stars and glowing gas.

The pressure within the gas balances the effect of gravitational pull between atoms so that contraction stops, and a stable state is achieved.

Stars can remain in this stable state for millions or billions of years. Eventually they run out of hydrogen gas to burn, causing the balance with gravity to fail and renewed contraction.

A supernova, a massive explosion of energy emitting the light equivalent to a whole galaxy, may mark the end of a star's life. During the explosion heavy elements may be formed and spewed out into the region between the stars, thus polluting the hydrogen and helium with the sort of material from which dust, rocks and planets are formed. This material will form part of future generations of stars. The core of the star may survive the supernova explosion as a neutron star (perhaps a pulsar) or even a black hole, which will trap light within what is called an event horizon, though still exerting a gravitational pull on external objects.

LIFE ON EARTH

OUR SUN is a second- or third-generation star, with its origins in another star's supernova detritus, with the planets created from the same material. As these smaller rocks cooled, gases were emitted from the rock, creating primitive, toxic atmospheres which, in the case of the earth, were sufficient to enable very low-level life, macromolecules, to assert themselves in the hot, uncomfortable oceans. With adaptation and survival, some of these molecules developed further, absorbing the hydrogen sulphide in the atmosphere, releasing oxygen (much like plants now release oxygen, but absorb carbon dioxide) and, on earth, ultimately created an atmosphere where certain sorts of fish, then reptiles, then mammals, then humans could survive.

It is entirely possible, under this model, that life on other planets could exist, created by chance combinations of molecules which adapted to their atmospheres and slowly altered their environment to suit their survival.

NASA photo of our sun with a spectacular flare.

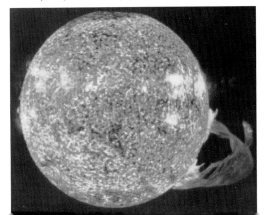

The Big Bang

The nature of the Big Bang remains a puzzle. Its occurrence is not in doubt, but its precise conditions present significant questions about the nature of time and leave room for religious and philosophical enquiry.

THE BIG BANG occurred between ten and twenty thousand million years ago. Soon after the Big Bang, the universe was much smaller and more compact than it is today, and immensely hot: in excess of fifteen hundred million degrees Celsius; so hot, in fact, that the universe consisted of pure radiation, whereas now, as we have seen, our current universe has a mixture of matter, in the form of stars, galaxies and a background radiation (see p. 166).

Modern theoretical physicists can predict back, with some confidence, to within a fraction of a second after the Big Bang, but earlier than this quantum mechanics and Einstein's General Relativity – our best theory of gravity – are needed together and as yet no satisfactory theory of quantum gravity has been produced. Ignoring such complications, at the point of a simplified Big Bang, the universe would have had infinite density and been infinitely hot. Such conditions can certainly not be handled by current theory.

Some theories regard the Big Bang as the beginning of a new cycle of expansion and contraction, that the Big Bang represents a simultaneous big crunch where the universe, having expanded for millions of years, has contracted only to expand again. There is a phenomenon called bulk viscosity, which refers to the slight increase in proportion of photons to nuclear particles created by the friction of such expansion and contraction: it still indicates that there must have been a beginning of some sort, even if there were several cycles of expansion and contraction.

This strays into areas of theoretical detail beyond real meaning to most of us, but the existence of an explosive beginning of time does have profound significance, not least for those of a religious temperament. Although literal-minded religious believers would reject this out of hand, it is significant that in 1951 the Catholic Church officially acknowledged the Big Bang as consistent with the meaning in *Genesis*.

Revealed in December 1995, this is the deepest-ever view from the NASA Hubble Space Telescope. It shows several hundred galaxies which have never been seen before, considered to be representative of the density and pattern of the universe in all directions. Some galaxies here were formed less than a thousand million years after the Big Bang.

Key Discoveries:
Time and Gravity

Key discoveries in the first decades of the twentieth century led to a radical new understanding of time and space, which moved science away from Newton's classical approach to more all-embracing theories of matter.

SIR ISAAC NEWTON, the founder of modern classical physics with his laws governing motion, light and gravity, posed a problem which his own theories could not resolve: if the universe were finite with matter distributed evenly (and undergoing no expansion), it would fall into a centre through gravity; if infinite, then a series of clumps of matter could form stars and the sun. His own physical theories led to the conclusion that the universe was finite, whereas, of course, there are stars and galaxies, suggesting that the universe could be infinite. As so often in science, grand theories present, for the moment of publication, theoretically complete explanations which push at the limits of our understanding, but the creator or discoverer of the theory knows better than anyone else where faults exist at the edges of their work. The same is the case for Einstein and for all the great twentieth-century discoveries which have progressively tackled the issues

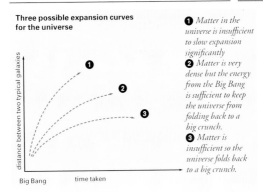

Three possible expansion curves for the universe

distance between two typical galaxies

Big Bang   time taken

❶ *Matter in the universe is insufficient to slow expansion significantly*
❷ *Matter is very dense but the energy from the Big Bang is sufficient to keep the universe from folding back to a big crunch.*
❸ *Matter is insufficient so the universe folds back to a big crunch.*

of infinity and gravity. The element of doubt has always been a prime motivating factor in the pursuit of scientific inquiry.

LIGHT AND SPECIAL RELATIVITY

IN THE last two decades of the nineteenth century, two discoveries led to a more complete understanding of time, motion and the effects of gravity. In 1865, James Clerk Maxwell demonstrated that all electromagnetic waveforms travel at fixed speeds and in 1887, Albert Michelson and Edwin Morley showed that light travelled at the same, absolute, rate in every direction, independent of the velocity of its source. Distance became measurable by using the speed of light as a constant; the distance from the earth to the moon can be monitored accurately by measuring the time taken for a laser pulse to bounce back from a corner-reflector left by astronauts.

These discoveries are amongst those which led to Einstein's first theory – his Special Theory of Relativity – which predicted that this absolute rate demonstrated that the speed of light was not dependent on the motion of an observer or an object, but rather that it is always the same (we now measure light as travelling 186,000 miles/300,000 km per second).

Einstein's equation $E=mc^2$ shows that the mass and total energy of a body are equivalent: a body's energy of motion

The four dimensions of space-time showing the three coordinates of space (in distance) and one of time

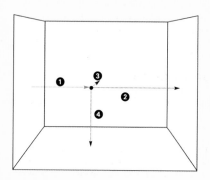

❶ *Time taken to reach this point in the room from the left wall*
❷ *Distance to the opposite wall*
❸ *Distance to the back wall*
❹ *Distance to the floor*

increases its mass. When a particle like an electron approaches the speed of light, its mass increases and it is theoretically impossible for it to reach the speed of light because this would require infinite energy.

Albert Einstein

Special Relativity shows that time must be considered in relation to space: an event or point can be described by using four co-ordinates: one of time and the three dimensions of space (see diagram) and the co-ordinates of such space-time are broadly interchangeable.

This space-time model forms the basis of modern science's theoretical explorations into the existence of a beginning and an end of time.

EINSTEIN'S GENERAL THEORY OF RELATIVITY

SPECIAL RELATIVITY did not take account of gravity and it was not until 1915 that Einstein discovered that a theory of gravity could be reconciled with Relativity. The result was his General Theory of Relativity, in which he showed the effects of gravity to be a consequence of space-time being warped by the mass and distribution of energy within it. Einstein's theory achieved the same general results as Newton's laws, but he was also able to account for some anomalies relating to the bending rays of light and the effect of particularly heavy bodies such as the sun. He showed that although our normal three-dimensional perception of the rotation of a planet around a star (rotating at a particular distance due to the gravitational pull of the star and the

The Geodesic Line

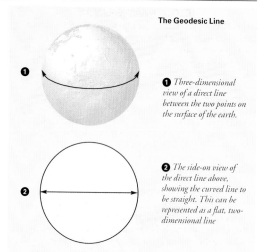

❶ *Three-dimensional view of a direct line between the two points on the surface of the earth.*

❷ *The side-on view of the direct line above, showing the curved line to be straight. This can be represented as a flat, two-dimensional line*

relative mass of the two bodies) could be observed as a circle or ellipse, it should be described as a geodesic line in a curved space-time continuum.

This different way of explaining gravity finally accounted for the dismissal of an absolute concept of time which had persisted from Aristotle to Newton and the absolute concept of space which both Galileo and Newton had discarded from Aristotle, yet still fully explained the gravitational effects within space and time.

CLOCKS AND GRAVITY

GENERAL RELATIVITY predicts that time will be slower, the closer it comes to a heavy mass like the earth, and that this gravitational pull has an effect on our personal experience of time. A clock placed at sea level will be observed to be slightly slower from the position of another, identical clock placed on top of a mountain. This observation shows a particular significance of the abandonment of absolute time, demonstrating that we have to agree a collective measurement of time to function as a society: the adoption of the most recent time standards (see p. 93) in the mid-1960s is based on the vibrations of the caesium atom which can be measured in millionths of a second and which operate independently of the motions of the stars.

RELATIVITY AND EXPANSION

IN UNIFYING a theory of gravity with Relativity, Einstein's new approach to space and time made predictions of an expanding universe with the possibility of an infinity of progress, by showing the effect of gravity on mass and demonstrating that the force of original propulsion from a Big Bang still provides sufficient momentum to push the universe outwards.

To do this, as Newton's laws also predicted (although not generally realised before the twentieth century), the mass of the universe must be below a certain level to ensure that expansion continues. It is speculated that the total matter of the universe, plus some unobservable dark matter, probably gives only 30 per cent of the total mass required to halt the current rate of expansion. If it did halt, the universe would theoretically collapse back on itself, taking as long to do so as it has taken so far to expand. The conditions for this happening are difficult to predict, but it has been suggested that our existence, as sentient, time-bound beings would not be possible in a contracting universe.

GRAVITY AND ESCAPE

THE ENERGY required to keep the universe expanding rather than contracting, uses the same principle as that applied to the energy required by a rocket designed to escape the downward force of earth's gravity. The original spaceships sent into space by the Americans and Russians in the 1950s and 1960s were over 100 metres tall, 90 metres of which consisted of fuel to provide enough energy to escape the gravitational pull of the earth. Once free of gravity the majority of the rocket was discarded and minimal amounts of fuel were expended reaching the moon for the famous landing by Neil Armstrong and his crew in 1969.

TIME TRAVEL

SOME MATHEMATICIANS and philosophers have stretched their credibility by proposing that a highly unlikely but interesting side effect of Relativity is the possibility of time travel. As space-time is a four-dimensional entity, with time to some extent interchangeable with the three dimensions of space, the movement from one location to another could have the same magnitude in the time or space co-ordinates, (although most scientists would distinguish qualitatively different 'time-like' and 'space-like' intervals). For instance, the interval between Shakespeare's time and our time has the same magnitude, in a sense, as the interval between Shakespeare's Globe Theatre and a point on the star, Antares.

Apollo 16, Saturn V spaceship, lifting off 16 April 1972: NASA's eighth manned voyage to the moon.

Key Discoveries: Quantum Mechanics

Study of the activity of atomic particles allows us to investigate the conditions and activity of the early universe.

QUANTUM MECHANICS, unlike Einstein's Relativity, focuses on the behaviour of atoms. It tells us that any particle has what can be described as a quantum state, which is a combination of its position and velocity. It also tells us that although we can describe a series of probable outcomes from an experiment on a particle, we cannot describe what will definitely happen.

Quantum mechanics replaced classical mechanics in the mid 1920s, after a series of observations and equations constructed by a number of mathematicians and physicists, following on from Einstein's observations several years earlier. Theories on waveforms and particles, disturbance and probability developed separately by Paul Dirac, Erwin Schrödinger and Werner Heisenberg led to a presentation in

1927 which has subsequently formed the basis of almost all the technological advances this century. Super and semi-conductors, lasers, microchips (in domestic appliances and computers), radios, televisions, calculators and genetic engineering all owe their existence to quantum theory.

HEISENBERG'S UNCERTAINTY PRINCIPLE

ONE OF the most accessible routes to quantum theory was Heisenberg's Uncertainty Principle, which was devised to cope with the frustrating phenomenon that the most effective way of measuring the position and velocity of a particle – by using quanta of light – disturbs the very properties of velocity or the position in an unpredictable way. As we have seen, light has no mass but still has energy, and the closer the quanta of light allow us to locate the particle, the more they affect it, making it impossible to predict its behaviour.

THE EARLY UNIVERSE

THIS HAS profound implications for the way we are able to measure the motion of bodies, their place in space-time and the way we can view the early moments of the universe. This is significant because it is clear, as discussed earlier, that the first few minutes of the Big Bang have defined the extent, content and motion of the universe as it is today.

In those first seconds and minutes after the Big Bang, the universe had immense density and maintained extremely high temperatures, conditions which quantum mechanics is equipped to study because it can give us a picture of the formation and behaviour of fundamental particles.

The familiar microchip, found in computers, cars and fridges.

Key Discoveries:
The Expanding Universe

One of the key discoveries enabling us to understand the beginnings of the universe, and time, was the observation in 1929 by Edwin Hubble, that the universe was expanding uniformly in all directions. The universe must therefore have had a single point of origin at a definite time in the past.

THROUGH THE evidence of Hubble's observations we now know that the universe is in a state of perpetual motion, with distant galaxies moving away from us in some cases at speeds close to the speed of light. Because this is not perceptible to the human eye, it took close observation of distant stars, the technology available to observe them (in the form of the brand new 100-in Mount Wilson telescope) and an understanding of the behaviour of light, to reveal this motion. For instance, observations with the Mount Wilson telescope revealed that a galaxy in Bootes at a distance of about two thousand million light years is moving away from us at a speed of nearly 40,000 kilometres per second. Such observations were impossible before the development of giant telescopes.

Later calculations by, amongst others, Howard Robertson and Arthur Walker in 1935, show that the rate at which the universe is expanding may be slowing down (leading to speculation that eventually it will stop and reverse its

Two million light years away, the Andromeda Galaxy, with two companion galaxies orbiting, is the nearest spiral galaxy to the Milky Way.

Stephen Hawking.

expansion, due to the force of gravity). Extrapolation from this model, which is caused by the relative density of matter and its gravitational effect, has enabled scientists like Nobel prize winner Steven Weinberg and, later, Stephen Hawking to calculate the date of the Big Bang being as between ten and twenty thousand million years ago.

REDSHIFT

HUBBLE HAD been studying what was thought to be a nebula, known as M31, within our own galaxy, when he discovered it was, in fact, another galaxy, by observing stars called Cepheid variables within it. We now know that our galaxy is one of one hundred thousand million, with each galaxy containing something like one hundred thousand million stars. Our galaxy is one hundred thousand light years across.

Hubble's later discoveries were achieved through the use of redshift as a method of measurement. Redshift uses the principles of the Doppler effect (commonly experienced in the different pitch of sound as a police car races towards, then past, a stationary pedestrian). As light is also a wave motion, with a frequency higher for blue light and lower for red light, in the same way, the observer of an approaching light source will see its colour shifted towards the blue, and the observer of a receding source will see its colour shifted towards the red. Hubble found that most galaxies, observed in every direction, had red-shifted spectra, with the more distant ones red-shifted more than the nearer ones. In fact we now know that the universe is expanding by 5–10 per cent every thousand million years.

View of the expanding universe, shown as a sphere

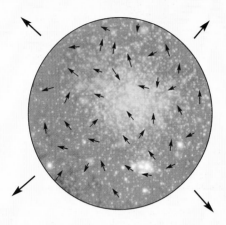

Hubble showed that the universe must be expanding in every direction; it grows like a balloon with the galaxies inside expanding simultaneously in all directions.

Key Discoveries:
Background Radiation

In 1965, observations of a regular radio noise were
made, leading to further predictions about
the date of the Big Bang and the beginnings
of the universe.

ON THE roof of the Bell Telephone Laboratories in New
Jersey, Arno Penzias and Robert Wilson measured a level
of radio noise above the minute electrical static of electrons.
They found that the noise could be measured as being
equivalent to the black-body radiation expected from an object
at a temperature of 3.5 Kelvin and that this radiation occurred
consistently in all directions. Their discovery could be
measured against the point of thermal equilibrium thought to
have pertained at a point within the first few minutes of the
Big Bang when the universe was too hot for matter to have
formed and therefore contained pure
radiation. It also enabled us to account
for the creation of the amount of
helium that exists in the universe,
which could only have been created at
temperatures high enough for protons
and neutrons to have combined.

*Penzias and Wilson in front of the horn
radio antennae at the Bell Laboratory in
New Jersey, USA.*

CONNECTING BACK TO THE EARLY UNIVERSE

THE OBSERVATION of the background radiation occurring in every direction, at every point in space, at the same consistent level, at a temperature which can be predicted back to the levels of temperature needed to produce the helium, gives us a means of reaching the origins of our universe. This electromagnetic radiation, in having no mass, travels at the speed of light and the radiation detectable today dates from the beginnings of our universe. Philosophically, in a sense we are able to both 'see' the beginnings of time and 'be' in our own time simultaneously.

Path of radiation from the Big Bang

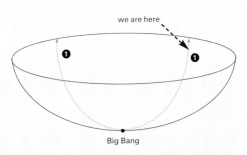

❶ *The universe is shown here as a sphere with two simple paths of radiation, expanding out from the Big Bang, tracing the overall expansion of the universe and showing how temperatures could be predicted back to conditions in the early universe.*

Light and Time

Philosophically, light, because of its immense speed
and lack of gravitational pull, carries a direct
connection to every point in the universe without
physically affecting any point in it.

A LIGHT YEAR is the distance travelled by light in an earth
year: 5.9 million million miles. Light from the sun takes
eight minutes to reach us, travelling at 186,000 miles per
second. According to Relativity, a body without mass, such as a
neutrino, can travel at the speed of light. A body with mass
would have to achieve infinite mass to reach the speed of light
but, as we have seen, this would need infinite energy so is
theoretically impossible.

CONNECTING TO ALL POINTS IN THE UNIVERSE

THE DISCOVERY of the constant background radiation
allows us, in Einstein's space-time model, to envisage the
beginnings of the universe and any other moment in time and
space being simultaneously encapsulated through the movement of a photon. To a photon, moving at the speed of light
means that it experiences the collapse of a star several thousand
million light years away from our sight of it several thousand
million years later at the same time. Light connects all points in

*Eta Carinae, a star one hundred times bigger than our sun is 8,000 light
years away from us. The dust clouds on either side conceal the star itself
which survived a massive explosion observed 150 years ago, but which
actually happened 8,000 years ago.*

the universe and in doing so connects all points in time and space, past, present and, arguably, future. In the beginning, the Christian, Islamic and Jewish sacred texts tell us that light filled the world and in a literal sense this is true, although it manifested itself in a way which is contrary to our own experience – as so-called 'black-body radiation' at a high temperature.

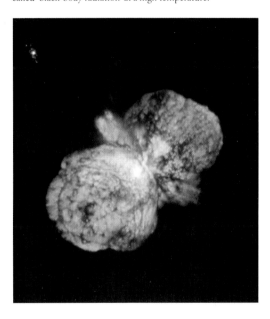

Arrows of Time

The simultaneous concurrence of all time and space, as explored in previous sections, does not preclude the existence of linear forms of time in our universe.

THE MODELS of mathematical physics have to allow for the phenomenon of linear time operating within the scope of the space-time continuum. Although, as we have seen, there are elements which enable us to 'see' the beginnings, the ends and the present of time (photons), there are combinations of elements – humans – which can only experience a narrow band of time past. Our psychological experience of time is that we feel time passing and remember the past, but do not remember the future. This experience can be measured using clocks or seasonally using the sun and moon, but it is subject to psychological variations of mood, slowing and speeding our sense of time, as explored earlier in this book (see p. 98).

Cosmologically, the progression of space from hot early density to expanded and expanding cooler and cooling immensity can be expressed in the terms of linear time.

Our experience of time in this way conforms to the Second Law of Thermodynamics, which says that ordered states always become disordered. This is the state of entropy which describes our experience of linear or 'real' time, showing that over the passage of time rocks break up and crumble, starting from a solid form and moving into disorder. The universe can be described in this way, starting as a single mass of high-density radiation, breaking up into an expanding varidensitied entity of variable masses, radiation and possible undiscovered dark matter.

The Milky Way, observed over time, and made from a number of composite images showing stars, which undergo their own lifecycle, their own arrow of time.

Can Time Begin?

Explicit in the familiar concept of arrows of time is a beginning. Scientists have tried to find a solution to notions of beginning in mathematical models, such as those of the hypersphere, which needs no start or end points with greater significance than any other points in the sphere.

A BEGINNING of time implies the existence of something before time. The notion of this as a possibility has preoccupied scientists and philosophers for generations. Aristotle believed that we cannot talk of a time before time, because this places it within the scope of time and becomes logically impossible. In modern scientific terms, as we have seen, we are unable to predict with certainty the events before the first hundredth second, but in mapping the extent, motion and mass of the universe, we have no mathematical equipment to measure or even speculate about events before the Big Bang. In these terms, the Big Bang is a real beginning, providing yet another of modern science's sophistries.

WHEN A BEGINNING IS NOT A BEGINNING

THERE ARE other ways of expressing the beginning of time without giving it the monumental importance of being the beginning of a giant arrow of time. In space-time it is possible to travel both backwards and forwards in time; as we have seen, light travels to all points in space and time, connecting time past with time future, so that a beginning is simply another event in the hypersphere.

One current mathematical model describes the universe as a closed system, where there is no point before the Big Bang, therefore there is no beginning as such. It has both the properties of infinity and of a boundry: it can operate like a giant three-dimensional Möbius strip (where two ends of a ribbon are twisted and joined to create a single surface) – there are no boundaries, but space-time keeps expanding infinitely.

In this model, the corpus of space-time has existence as itself, unaffected by any other forces. This provides a system where the potential of time is more important than the actuality of time because the potential could be manifested in its entirety, at once. Aristotle explored the notion of stasis as one of potential, where change and motion provided the actuality of time.

This emphasis on stillness, which can be expressed either as a still point or a system of still potential, leaves us with some critical questions: why did the Big Bang occur when it did and what was the agent of change? Religious enquiry might suggest that a god plays the significant role here.

Möbius strip

two ends joined here

one twist

A strip of paper or cloth is twisted once and the ends joined to create a single surface where a body can travel both in and outside the loop.

God and the State of Being

Modern science is popularly thought to have ruled out the need for a god. In fact, the uncertainty principles behind quantum mechanics and the enquiries about the beginning of time, can be interpreted as allowing for a motivating force which some might call God.

ANCIENT GREEK philosophers such as Plato and Aristotle are very instructive in discussions where religion and science converge. As the basis of most western philosophy they have informed the nature of discussion about our world and our place in it. As pre-Christians, they address, along with Jewish philosophers of the time, the great questions which were assimilated into the new belief systems of Christianity when it emerged.

Plato, in *Timaeus* said, 'there were no days, nights, months and years before the heavens came into being, but he devised that they should come into being just when the heavens were assembled. These are all parts of time, and "was" and "shall be" are forms of time which have come into being. We are wrong when we apply them ... to the everlasting being ... really only "is" belongs properly to it'.

Plato would have been perfectly comfortable with Einstein's concepts of space-time, but here he also

'The Fall of the Giants', a late Renaissance 'trompe l'oile' painting.

addresses the notion of a god which 'is': a force which is both in time and encompasses it. (Incidentally, he also dealt with time as 'the object of opinion and irrational sensation, coming to be and ceasing to be, but never fully real', where our normal experience of time is not real in the bigger universe of 'being'.)

If we look at *Exodus* in the Bible we see Moses being addressed by God, 'I am who I am, Say I am has sent me to you.' In old Hebrew terms *Yahweh* is the word for God – later shifting linguistically into Jehovah – meaning 'he that is'. Again we see the emphasis on being, on existing, of encompassing not participating within the confines of time.

The fourth-century St Augustine brought much of Plato's thinking into Christianity, providing it with much muscular philosophical thought: 'say that infinite things are past God's knowledge ... say that God knows not all numbers.... What madman would say so? What are we mean wretches to limit his knowledge?'

Dante, in his *Paradiso*, finds himself led through space, through the spheres of angels to a point called Empyrean, where God resides. In this perspective God exists at a still point, with the universe in the form of angels, then the planets, radiating outwards, as though from a singularity at the point of the Big Bang. The perspective is the same as the sliced view through the mathematical hypersphere.

We can see that God as an infinite being with omniscience, encompassing the whole of time, is not necessarily inconsistent with modern scientific ideas about space-time.

Dante views the Empyrean, and peers into the eye of the universe (see image on p. 13).

Infinity

Like the uncertainty principle, the struggle to define
and acknowledge infinity is undermined by the
methods of doing so which mark boundaries and
therefore turn the infinite into a measurable finitude.

INQUIRIES INTO infinity have existed since philosophers
started thinking and astronomers started observing the skies.
Infinity brushes against all the questions about the beginning
of time, the existence of God, about our experience of the
passage of time against the actual curved state of the universe.

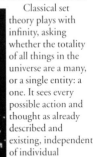

Classical set
theory plays with
infinity, asking
whether the totality
of all things in the
universe are a many,
or a single entity: a
one. It sees every
possible action and
thought as already
described and
existing, independent
of individual

*The spiral galaxy NGC
6946. Will such galaxies,
always exist, infinitely
creating new stars, or will
they eventually stop?*

motivation, waiting to be discovered and revealed. For the set theorist, Einstein's Theory of Relativity was waiting to be discovered by Einstein, Leonardo da Vinci's *Mona Lisa* was waiting to be painted. This is consistent with other theories on the nature of infinity, with both Newton and Einstein, for instance, believing that time could be experienced in scientific terms, both forwards and in reverse. The space-time concept of the Special Theory of Relativity would particularly support this concept.

It is also possible to construct models of infinity which include endless cycles of expansion and contraction, all curving back to the same singular point in hyperspace.

An infinity of expanding and contracting universes

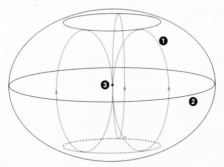

Space-time model showing a theoretical representation of endless cycles (three shown) of expanding and contracting universes, starting at the point of the Big Bang, ❸, expanding in space, ❷, over time, ❶, then folding back to ❸ in a big crunch.

INFINITY AND NUMBERS

IN ZEN Buddhism *prajna* knowledge is intuitive and holistic, capable of grasping the nature of an infinity of numbers, without making a finitude from it. *Vijnana* knowledge is the rational thought where the counting of numbers from 1, 2, 3, etc. can go on forever. Numbers provide an interesting insight into infinity. Pythagoras and his followers placed supreme importance in mathematical form as the ultimate reality, with everything being described by mathematics. Numbers were regarded as the ruling force, with 10 being perfect because it was the sum of the four primary numbers 1+2+3+4. Mathematically, numbers can be subdivided infinitely, as can events in space-time, which can be subdivided into instants, which can be theoretically subdivided infinitely.

A UNIFIED THEORY

THE MATHEMATICAL models of space-time help us to look beyond the apparent physical realities which show that particles can only be broken down into their components. Inevitably, there is a clear difference between mathematical models, physical evidence and psychological experience of the conditions of infinity.

As we start a third millennium of the western world, philosophical inquiry, mathematical physics and pure mathematics are furiously searching for a unified theory of gravity. It would combine the effects of Relativity and quantum mechanics and enable us to predict all motion from the point of the Big Bang into infinity, which could be defined without undermining our many and varied notions of time.

From the naked eye to observatory telescopes, mankind continues to search the skies for answers to life, space and time.

Glossary

Absolute Time: Newton's concept of time which flows through space at a uniform rate whatever the starting point; people at different points experience the same sense of the present.

Arc degrees: Using the angle of the surface of the earth, it is possible to measure time. 15° of arc is one hour of time, one degree is four minutes.

Astrolabe: Early astronomical instrument designed to measure the altitude of the stars and planets; used extensively at sea in the Middle Ages.

Atom: Unit of matter with a nucleus (with a positive electrical charge) orbited by electrons (with a negative charge); the nucleus contains protons and neutrons.

Atomic clock: Clock which keeps time by using natural frequencies of atoms; very accurate and stable because they are not affected by temperature, pressure or humidity.

Atomic second: The second is defined as 9,192,631,770 oscillations of a caesium atom and is now the international standard of time measurement, because the atom is not subject to change due to temperature, motion, pressure or humidity.

Autumnal equinox: Usually around 22 September, the point when the sun moves across the equator from northern to southern hemispheres. Day and night are of equal length.

Balance wheel: Part of the mechanism which regulates the beat of a timepiece.

Big Bang: Supposed by many theoretical scientists to be the beginnings of the universe as we know it, containing very high temperatures and very dense concentrations of matter which exploded, expanding outwards to form the universe.

Black Hole: Object of massive density and gravitational force from which even light cannot escape. Probably formed from a collapsed star, and may have similarities with the state of the early universe.

Carolingian: Named after the Carolingian Franks who ruled in France AD 751–987 and Germany to AD 911, a time of great cultural development in the West.

Chronometer: Robust timepiece designed to maintain accuracy in stressful conditions (high or low temperatures or pressure), particularly used at sea.

Classical Mechanics: Study of motion and position of particles, as defined by Newton.

Coordinated Universal Time: Introduced in 1972 to replace Greenwich Mean Time as the international standard; based on the atomic second, which is more accurate than reliance on the slightly irregular motion of the earth around the sun.

Cosmology: Study of the nature and origins of the universe.

Doppler effect: Change in frequency of a signal, as caused by the relative motion of the source and the observer.

Easter: Christian Feast celebrating the risen Christ; based on the date for the Jewish Passover, which marks the first full moon of Spring. The Christian Easter always falls on the first Sunday after the first full moon after the Spring equinox.

Electron: Elementary particle which possesses a very light mass. The interactions between electrons and atomic nuclei cause the formation of all atoms and molecules.

Entropy: In relation to this book, a lack of order or pattern, the degeneration of order into chaos.

Ephemeris Time: Measured by tables which predict the future astronomical positions of the sun, moon and other stars.

Equation of Time: Difference between mean and solar time, showing the position of the sun at the Meridian

Escapement: Mechanism in a timepiece which consists of a toothed wheel and anchor, designed to provide a regular pulse of energy to a balance or pendulum.

Event horizon: Boundary of a black hole, from which light cannot escape.

Event: Point in space-time which can be described by its three co-ordinates of space and one of time

Frequency: Number of cycles per second of a waveform.

Fusee: Type of escapement which regulates the flow of energy from the spring to the hands of a clock.

General Relativity: Einstein's second great theory of matter, time and space; he unified his Special Theory of Relativity with laws of gravity to explain how the laws of physics should be the same for all bodies, regardless of how they are moving in relation to each other. Gravity is described as a curvature in space-time.

Geocentric: Describing a model of the solar system where all the planets revolve around a stationary earth. The moon has a geocentric orbit.

Geodesic: Shortest path between two points on a surface. A geodesic on sphere has to go round the surface of a sphere.

Greenwich Mean Time: Local time at Greenwich in London which sits on the meridian line and was used as the international standard for time from the mid-1800s until 1972. Based on observations from the Greenwich Meridian the 24-hour cycles of motions of the earth against the stars.

Gregorian Calendar: Measured the year as 365.2425 days and introduced by Pope Gregory XIII in 1582; finally adopted in Britain in 1752.

Heliocentric: Describing a model of the solar system where all the planets revolve around the sun. The earth has a heliocentric orbit.

Hypersphere: Mathematical model which contains more than three dimensions. In the context of theoretical physics, it is a way of describing the four dimensions of space and time.

Intercalation: Unit of time, commonly a second, day or month, inserted into the calendar to align with the agreed national or international standard.

Julian Calendar: Measured the year as 365.25 days; introduced by Julius

Caesar in 46 BC. Superseded by the more accurate Gregorian Calendar.

Kelvin: Temperature scale which uses absolute zero as the start of its scale. Centigrade uses the melting point of ice as its zero, in Kelvin this point is 273.15 K.

Latitude: Distance in degrees north or south of the earth's equator.

Leap Second: Added in 1972 and inserted almost every year, it ensures that the difference between Universal Coordinated Time and earth-based time like GMT (used by ships which rely on astronomical observations for their time) is never more than 0.9 seconds.

Leap Year: Year in which February has 29 days rather than 28, added as an intercalation.

Light Year: Distance that light travels in one year at 186,000 miles/300,000 km per second.

Longitude: Distance in degrees east or west of the earth's prime meridian (the Greenwich Meridian).

Lunar month: Time the moon takes to travel around the earth. Also known as the synodic month, its length is currently 29.5305889 days; in 2010 it will be 29.5305891 days.

Maser clock: Maser stands for Microwave Amplification by

Stimulated Emission of Radiation. A maser clock amplifies oscillations of the caesium atom to ensure consistent mensurability to give accurate time data.

Mass: In physics, the measurement of the matter of a body, expressed either as its resistance to acceleration or the force experienced due to a gravitational field.

Master clock: Caesium clocks, run by government institutions, used to measure time independently of each other so that an agreed standard can be used internationally. International Bureau of Weights and Measures in Paris coordinates the measurements and international time standards.

Mean Time: Also called clock time. A uniform standard which averages out the slight variations of the motion of the sun through the year.

Meridian: Imaginary lines connecting the North and South Poles which mark the longitude of the earth. Greenwich, by an international agreement in 1884 lies on the prime meridian, 0 .

Metonic Cycle: Named after Meton, a fifth century BC Greek astronomer. In a single year the 12 lunar months make a shorter year than the measurement of the tropical year; Meton observed that the tropical year coincides with a definite number of lunar months

every 19 years (actually 234.997 months are equal to 19 years).

Millennium: Cycle of 1000 years. First millennium began on year one AD (there was no year zero) ending in 1000. Second millennium began in 1001, ending, contrary to popular understanding, in 2001. Some argue that the first millennium began at the point before year one, ending in 999, so 2000 is the millennium year.

Molecule: The simplest unit of a chemical compound that can exist, consisting of two or more atoms held together by chemical bonds.

Nocturnal: Early device, used particularly in the Middle Ages, for describing the time at night.

Photon: Particle of light. Light has no matter but carries energy in the form of electromagnetic radiation.

Quantum mechanics: Superseded Classical mechanics in 1920s as the method to describe and predict the behaviour of microscopic systems, expressing the motion and position of a particle as a single entity, a quantum state.

Red Shift: Observation of the reddening of light emitted from a star which is moving away from an observer; caused by the Doppler effect.

Rest energy: The energy in a particle which would be released if the mass of the particle was destroyed. Predicted by Einstein's $E=mc^2$.

Sidereal Day: Time taken by the earth on one turn of its axis, as measured against a fixed star.

Sidereal Time: Time measured by the passage of the stars, giving a year of 365.256366 days.

Singularity: A point, such as a black hole, or at the moment of Big Bang, where the predictions of the laws of physics cannot fully explain the activity or nature of its existence; they can be described as having infinite curvature in space-time.

Solar Time: Time measured by the passage of the sun across the sky.

Space-time: Concept, developed by Einstein, where the dimension of time has the same value as the three dimensions of space.

Special Relativity: Einstein's first great theory of matter, time and space. Predicts the laws of physics for bodies which move at constant speeds relative to each other, with the speed of light remaining constant.

Summer solstice: Longest day in the northern hemisphere, usually 21 June, when the sun reaches its northernmost latitude.

Thermodynamics: Science of heat and work.

Time dilation: Phenomenon of time slowing down, either by an increase in speed or the proximity of a large gravitational field (such as from a planet).

Tropical Year: Measurement of a year from a fixed point. Current length is 365.242190 days. It does vary slightly though, in 2010 it will be 365.242184 days.

Verge and Foliot: Early type of escapement; regulates the flow of energy to the hands of a clock mechanism.

Vernal equinox: Usually around 20 March, the point when the sun is moving across the equator from the southern to northern hemispheres; day and night are of equal length.

Wavelength: Distance between two peaks or troughs of a wave, such as radio waves or light radiation.

Winter solstice: Shortest day, when the sun reaches its southernmost latitude. Usually around 21 December in the northern hemisphere.

Acknowledgements

The author would like to express his grateful thanks to Sonya Newland for her support and encouragement in the writing of this book. Thanks also to Josephine Cutts, Claire Dashwood and Frances Banfield. The book is dedicated to Douglas Wells (so full of life, no sense of time).

All graphics (15, 23, 31, 36, 46, 61, 73, 76, 82, 93, 95, 153, 154, 158, 165, 167, 173, 177) and photo manipulations (13, 103, 108, 111, 179) are courtesy of Foundry Arts.

Picture credits: Allsport: pp. 118-119, 120(t), 121; The Bridgeman Art Library: pp. 174; The Worshipful Company of Clockmakers' Collection, UK: pp. 77; The British Art Library: pp. 43; Christie's Images: pp. 5, 107, 112, 115; (c) Copyright 1999, The Nasdaq Stock Market, Inc. Reprinted with the permission of The Nasdaq-Amex Market Group: pp. 111; (c) Crown Copyright 1998 Reproduced by permission of the controller of HMSO; The Dean and Chapter of Salisbury/ Steve Day: pp. 71; The Dean and Chapter of Wells: pp. 72; Foundry Arts: pp. 62(r), 127, 134, 136, 137, 139, 175 Alex Courtney: pp. 98, Karen Villabona:

pp. 117; Greg Evans International: pp. 101, 103(all), 104, 106(all), 111; Image Select: pp. 24; Image Select/Ann Ronan: pp. 26, 89(t), 92; Image Select/C.F.C.L.: pp. 22, 110; Image Select/Giraudon: pp. 26-27, 50; Lucent Technologies Bell Labs Innovations: pp. 166; Mary Evans Picture Library: pp. 6, 1117, 23, 31, 33(all), 35, 38, 39, 40, 41, 42, 44, 45, 47, 49, 56, 90(all); 120(b), 125(all), 135, 138, 179; NASA: pp. 13, 143(all)145, 146, 148-149150, 163, 169, 171, 176; National Maritime Museum Picture Library: pp. 68, 78, 79(t), 79(b), 84, 88(all), 91, 94, 96-97(all); Parliamentary Copyright 1995: pp. 116; Peter Sanders Photography: pp. 19; Photo: AKG London/Erich Lessing: pp. 124; Popperfoto: pp. 63, 69, 99, 109, 138(b); Punch Ltd: pp. 105; Redferns: pp. 129; Ronald Grant Archive: pp. 130-131(all), 140-141; Science Museum/Science & Society Picture Library: pp. 7, 25, 52-53, 58, 60, 62, 6465, 66-67, 74-75, 80-81, 83, 85, 86, 89(b), 95(b), 100, 108, 152, 159, 164; Still Pictures/ Y Noto Campanella: pp. 113, Richard J. Wainscoat 179; (c) Tate Gallery, London 1999: pp. 122-123, 132-133; Topham: pp. 18, 21, 23(r), 30, 37, 54-55, 102, 126, 128, 160.

Bibliography/Further Reading

A number of original and secondary sources have been referred to in the main parts of this book, but for those interested further the list below provides some pointers for additional research. The Internet is also an invaluable resource, but should be treated with some care because the general fascination with time has led to many interested but unqualified authors placing dubious information on the net.

Time, The Calendar and General Cultural Background

Borst, Arno, *The Ordering of Time: From the Ancient Computus to the Modern Computer*, Polity Press, Oxford

Duncan, David Ewing, *The Calendar*, 4th Estate, London

Eliade, Mercea, *A History of Religious Ideas*, University of Chicago

Eliade, Mercea, *The Sacred and Profane: The Nature of Religion*, Harcourt Brace, New York

Grant, Edward, *The Foundations of Modern Science in the Middle Ages*, Cambridge University Press

James, William, *Varieties of Religious Experience*

Kant, Immanuel, *Universal Natural History and Theory of the Heavens*,

Kline, Morris, *Mathematical Thought from Ancient to Modern Times*, OUP

Landes, David, *Bankers and Pashas: International Finance and Economic imperialism in Egypt*, Harvard University Press

Le Goff, Jacques, *Time Work and Culture in the Middle Ages*, University of Chicago Press

Ling, T., *A History of Religion East and West*, Macmillan, London

McNeill, Waldman, *The Islamic World*, University of Chicago Press

Needham, Joseph, *Science and Civilisation in China*, Cambridge University Press

Penguin Atlas of World History

Rahman, Fazlur, *Islam and Modernity*, University of Chicago Press

Smart, Ninian, *The Religious Experience of Mankind*, Fontana Press, London

Times Atlas of Ancient History

Watts, Alan, *The Way of Zen*, Penguin, London

White Jnr, Lynn, *Medieval Technology and Social Change*, OUP

Time and Clocks

Baillie, G. H., *Watchmakers and Clockmakers of the World*, NAG Press

Burton, Eric, *The History of Clocks and Watches*, Rizzoli, New York

Camerer Cuss, T. P., *The Camerer Cuss Book of Antique Watches*, Antique Collectors Club, Suffolk

Edwardes, Ernest and J. Sherratt, *The Story of the Pendulum Clock*

Goode, Richard, *Britten's Watch and Clockmaker's Handbook, Dictionary and Guide*, 16th ed., Methuen, London

Landes, David, *Revolution in Time*, Harvard University Press

Price, *Heavenly Clockwork: the Great Astronomical Clocks of Medieval China*

Royal Observatory, Greenwich, Royal Maritime Museum

Walker, Carlo Cipolla, *Clocks and Culture*, New York

Time and Navigation

Betts, Jonathan, *Harrison*, National Maritime Museum, London

Cotter, Charles, Hollis and Carter, *A History of Nautical Astronomy*, Toronto

Gould, Rupert, *The Marine Chronometer*, Holland Press, London

Howse, Derek, *General Cultural Background Greenwich Time and the Discovery of the Longitude*, OUP

Sobel, Dava, *Longitude*, 4th Estate, London

Time and Behaviour

Moore-Ede, Sulzman, Fuller, *The Clocks That Time Us: Physiology of the Circadian Timing System*, Harvard University Press

Winfree, A., *The Geometry of Biological Time*, Springer Verlag

Time and Space

Calder, N., *Einstein's Universe*, BBC

Cornell and Gorenstein, *Astronomy from Space*, MIT, Massachussetts

Correspondence of Isaac Newton, The Cambridge University Press

Gribbin, John, *In Search of Schödinger's Cat*, Black Swan

Hawking, Stephen, *A Brief History of Time*, Bantam Press

Hawking, Stephen and G.F.R. Ellis, *The Large Scale Structure of Space-Time*, Cambridge University Press

Hoyle, Fred, *Astronomy and Cosmology – A Modern Course*, W. H. Freeman & Co

Hubble, Edwin, *The Realm of the Nebulae*, Yale University Press

Penrose, Roger, *The Emperor's New Mind*, OUP

Weinberg, Steven, *The First Three Minutes*, Flamingo

General Scientific Background

Allen, C. W., *Astrophysical Quantities*, Athlone Press, London.

Brooks, D. and E. Wiley, *Evolution as Entropy: Towards a Unified Theory of Biology*, University of Chicago Press

Cohen, Paul J., *Set Theory and the Continuum Hypothesis*, Benjamin

Koyre, Alexandre, *From the Closed World to the Infinite Universe*, John Hopkins Press, Balimore

Seielstad, George, *Cosmic Ecology*, University of California Press

Stewart, I., *Does God Play Dice? The Mathematics of Chaos*, Blackwell

Time and Philosophy

Coveney, Peter and Roger Highfield, *The Arrow of Time*, Flamingo

Kant, Immanuel, *Critique of Pure Reason*

Rucker, Rudy, *Infinity and the Mind*, Paladin

Solomon and Higgins, *Short History of Philosophy*, OUP

Sorabji Richard, *Time, Creation and Continuum*, Duckworth, London

Other Interesting or Useful books

Gould, Stephen Jay, *Time's Arrows, Time's Cycle,* Penguin

Gunn, James, *The Road to Science Fiction,* NEL, London

Morris, R., *The Dark Abyss of Time,* University of Chicago Press

Sheldrake, Rupert, *The Presence of the Past,* London

At the time of going to press the core websites worth exploring on the Internet are: the Greenwich Maritime Museum/Royal Observatory website at www.greenwich2000.com/time; the US equivalent of the Greenwich site is physics.nist.gov/GenInt/Time/time.html

Other useful sites include: www.maa.mhn.de/Scholar/times.html, www.fordham. edu/halsall/ sbook.html, www.nasa.gov.

Useful Addresses and Organisations

Antiquarian Horological Society
New House, High Street
Ticehurst
Wadhurst , East Sussex, TN5 7AL

British Horological Institute
Upton Hall
Upton
Newark, Notts, NG23 5TB
Tel: 01636 813795

Clockmakers Company Museum
Guild Hall Library
Aldermanbury
London EC2P 2EJ
Tel: 0171 606 3030

Musée International d'Horlogerie
Rue des Musees 29
La Chaux-de-Fonds
CH-2301, Switzerland
Tel: 00 41 32 9676861

Musée de d'Horlogerie,
15, Route de Malagnou
CH-120B
Geneve, Switzerland
Tel: 00 41 22 4186470

Science Museum
Exhibition Road
South Kensington,
London SW7 2DD
Tel: 0171 938 8000

Smithsonian Institution
2 Constitution Ave.
American History Museum
Washington 20560, USA
Tel: 001 202 357 3029

Index

COLLINS GEM
BABIES' names

COLLINS GEM
BEER

COLLINS GEM
BIRDS

COLLINS GEM
CALORIE Counter

COLLINS GEM
FACT FILE

COLLINS GEM
FENG SHUI

COLLINS GEM
F.LAGS

COLLINS GEM
Healthy EATING

COLLINS GEM
QUOTATIONS

COLLINS GEM
SAS Self-Defence

COLLINS GEM
SAS Survival Guide

COLLINS GEM
SEASHORE

COLLINS GEM
TREES

COLLINS GEM
Understanding DREAMS

COLLINS GEM
WILD flowers

COLLINS GEM
WINE Dictionary